Models of Horizontal Eye Movements

Part 3

A Neuron and Muscle Based Linear Saccade Model

Synthesis Lectures on Biomedical Engineering

Editor
John D. Enderle, *University of Connecticut*

Lectures in Biomedical Engineering will be comprised of 75- to 150-page publications on advanced and state-of-the-art topics that span the field of biomedical engineering, from the atom and molecule to large diagnostic equipment. Each lecture covers, for that topic, the fundamental principles in a unified manner, develops underlying concepts needed for sequential material, and progresses to more advanced topics. Computer software and multimedia, when appropriate and available, are included for simulation, computation, visualization and design. The authors selected to write the lectures are leading experts on the subject who have extensive background in theory, application and design. The series is designed to meet the demands of the 21st century technology and the rapid advancements in the all-encompassing field of biomedical engineering that includes biochemical processes, biomaterials, biomechanics, bioinstrumentation, physiological modeling, biosignal processing, bioinformatics, biocomplexity, medical and molecular imaging, rehabilitation engineering, biomimetic nano-electrokinetics, biosensors, biotechnology, clinical engineering, biomedical devices, drug discovery and delivery systems, tissue engineering, proteomics, functional genomics, and molecular and cellular engineering.

Models of Horizontal Eye Movements: Part 3, A Neuron and Muscle Based Linear Saccade Model
Alireza Ghahari and John D. Enderle
2014

Digital Image Processing for Ophthalmology: Detection and Modeling of Retinal Vascular Architecture
Faraz Oloumi, Rangaraj M. Rangayyan, and Anna L. Ells
2014

Biomedical Signals and Systems
Joseph V. Tranquillo
2013

Health Care Engineering, Part I: Clinical Engineering and Technology Management
Monique Frize
2013

Health Care Engineering, Part II: Research and Development in the Health Care
Environment
Monique Frize
2013

Computer-aided Detection of Architectural Distortion in Prior Mammograms of Interval
Cancer
Shantanu Banik, Rangaraj M. Rangayyan, and J.E. Leo Desautels
2013

Content-based Retrieval of Medical Images: Landmarking, Indexing, and Relevance
Feedback
Paulo Mazzoncini de Azevedo-Marques and Rangaraj Mandayam Rangayyan
2013

Chronobioengineering: Introduction to Biological Rhythms with Applications, Volume 1
Donald McEachron
2012

Medical Equipment Maintenance: Management and Oversight
Binseng Wang
2012

Fractal Analysis of Breast Masses in Mammograms
Thanh M. Cabral and Rangaraj M. Rangayyan
2012

Capstone Design Courses, Part II: Preparing Biomedical Engineers for the Real World
Jay R. Goldberg
2012

Ethics for Bioengineers
Monique Frize
2011

Computational Genomic Signatures
Ozkan Ufuk Nalbantoglu and Khalid Sayood
2011

Digital Image Processing for Ophthalmology: Detection of the Optic Nerve Head
Xiaolu Zhu, Rangaraj M. Rangayyan, and Anna L. Ells
2011

Modeling and Analysis of Shape with Applications in Computer-Aided Diagnosis of
Breast Cancer
Denise Guliato and Rangaraj M. Rangayyan
2011

Basic Feedback Controls in Biomedicine
Charles S. Lessard
2009

Understanding Atrial Fibrillation: The Signal Processing Contribution, Part I
Luca Mainardi, Leif Sörnmo, and Sergio Cerutti
2008

Understanding Atrial Fibrillation: The Signal Processing Contribution, Part II
Luca Mainardi, Leif Sörnmo, and Sergio Cerutti
2008

Introductory Medical Imaging
A. A. Bharath
2008

Lung Sounds: An Advanced Signal Processing Perspective
Leontios J. Hadjileontiadis
2008

An Outline of Informational Genetics
Gérard Battail
2008

Neural Interfacing: Forging the Human-Machine Connection
Susanne D. Coates
2008

Quantitative Neurophysiology
Joseph V. Tranquillo
2008

Tremor: From Pathogenesis to Treatment
Giuliana Grimaldi and Mario Manto
2008

Introduction to Continuum Biomechanics
Kyriacos A. Athanasiou and Roman M. Natoli
2008

The Effects of Hypergravity and Microgravity on Biomedical Experiments
Thais Russomano, Gustavo Dalmarco, and Felipe Prehn Falcão
2008

A Biosystems Approach to Industrial Patient Monitoring and Diagnostic Devices
Gail Baura
2008

Artificial Organs
Gerald E. Miller
2006

Signal Processing of Random Physiological Signals
Charles S. Lessard
2006

Image and Signal Processing for Networked E-Health Applications
Ilias G. Maglogiannis, Kostas Karpouzis, and Manolis Wallace
2006

Models of Horizontal Eye Movements: Part 3, A Neuron and Muscle Based Linear Saccade Model
Alireza Ghahari and John D. Enderle

ISBN: 978-3-031-00533-6   paperback
ISBN: 978-3-031-01661-5   ebook

DOI 10.1007/978-3-031-01661-5

A Publication in the Springer series
SYNTHESIS LECTURES ON BIOMEDICAL ENGINEERING

Lecture #53
Series Editor: John D. Enderle, *University of Connecticut*
Series ISSN
Print 1930-0328   Electronic 1930-0336

Models of
Horizontal Eye Movements

Part 3

A Neuron and Muscle Based Linear Saccade Model

Alireza Ghahari
University of Connecticut

John D. Enderle
University of Connecticut

SYNTHESIS LECTURES ON BIOMEDICAL ENGINEERING #53

ABSTRACT

There are five different types of eye movements: saccades, smooth pursuit, vestibular ocular eye movements, optokinetic eye movements, and vergence eye movements. The purpose of this book series is focused primarily on mathematical models of the horizontal saccadic eye movement system and the smooth pursuit system, rather than on how visual information is processed. A saccade is a fast eye movement used to acquire a target by placing the image of the target on the fovea. Smooth pursuit is a slow eye movement used to track a target as it moves by keeping the target on the fovea. The vestibular ocular movement is used to keep the eyes on a target during brief head movements. The optokinetic eye movement is a combination of saccadic and slow eye movements that keeps a full-field image stable on the retina during sustained head rotation. Each of these movements is a conjugate eye movement, that is, movements of both eyes together driven by a common neural source. A vergence movement is a non-conjugate eye movement allowing the eyes to track targets as they come closer or farther away.

In Part 1, early models of saccades and smooth pursuit are presented. A number of oculomotor plant models are described therein beginning with the Westheimer model published in 1954, and up through our 1995 model involving a 4th-order oculomotor plant model. In Part 2, a 2009 version of a state-of-the-art model is presented for horizontal saccades that is 3rd-order and linear, and controlled by a physiologically based time-optimal neural network.

In this book, a multiscale model of the saccade system is presented, focusing on the neural network. Chapter 1 summarizes a whole muscle model of the oculomotor plant based on the 2009 3rd-order and linear, and controlled by a physiologically based time-optimal neural network. Chapter 2 presents a neural network model of biophysical neurons in the midbrain for controlling oculomotor muscles during horizontal human saccades. To investigate horizontal saccade dynamics, a neural circuitry, including omnipause neuron, premotor excitatory and inhibitory burst neurons, long lead burst neuron, tonic neuron, interneuron, abducens nucleus, and oculomotor nucleus, is developed. A generic neuron model serves as the basis to match the characteristics of each type of neuron in the neural network. We wish to express our thanks to William Pruehsner for drawing many of the illustrations in this book.

KEYWORDS

saccade, neural network, neural dynamic, neural model, neural input, compartmental approach, burst firing, system identification, glissadic overshoot, time-optimal control

Contents

Acknowledgments

We wish to express our thanks to William Pruehsner for drawing many of the illustrations in this book.

Alireza Ghahari and John D. Enderle
September 2014

CHAPTER 1

2009 Linear Homeomorphic Saccadic Eye Movement Model

1.1 INTRODUCTION

In 1988, Enderle and Wolfe[1] described using the system identification technique to estimate the parameters of a 4th order model of the oculomotor plant presented in Section 3.6 of Part 1, and the active-state tensions during the saccadic eye movement. The active-state tension is modeled as a low-pass filtered pulse-step waveform, as described earlier. Parameter estimates are calculated for the model using a conjugate gradient search program that minimizes the integral of the absolute value of the squared error between the model and the data in the frequency domain. Initial parameter estimates are based on physiological evidence. For saccades that end without post-saccade behavior, the estimation results are in excellent agreement with the data as shown in Figures 3.31–3.33.

In this chapter, we summarize the fast eye movement system, including normal saccades and those with a dynamic or a glissadic overshoot based on a model by Zhou et al. (2009). To analyze post-saccade behavior, the neural input to the muscles is now described by a pulse-slide-step of neural activity, supported by physiological evidence (Goldstein, 1983). The slide is a slow exponential transition from the pulse to the step. We will also explore system identification to estimate parameters for this model and a time-optimal controller.

1.2 OCULOMOTOR PLANT

The oculomotor plant is shown in Fig. 1.1. The eye muscles are identical to those in Section 1.2 of Part 2, where all elements are linear. This linear muscle model exhibits accurate nonlinear force-velocity and length-tension relationships. No other linear muscle model to date is capable of accurately reproducing these nonlinear relationships.

The inputs to the muscle model are the agonist and antagonist active-state tensions, which are derived from a low-pass filtering of the saccadic neural innervation signals. As previously mentioned, the neural innervation signals are typically characterized as a pulse-step signal, or

---

[1]Some of the material in this chapter is an expansion of a previously published paper: Zhou, W., Chen, X., and Enderle (2009). An Updated Time-Optimal 3rd-Order Linear Saccadic Eye Plant Model. *International Journal of Neural Systems*, Vol. 19, No. 5, 309–330 and the MS Thesis by Wei Zhou, A Mathematical Model For Horizontal Saccadic Eye Movement Based on a Truer Linear Muscle Model and Time-Optimal Controller, University of Connecticut, 2006. It is also a summary of Part 2 in this book series.

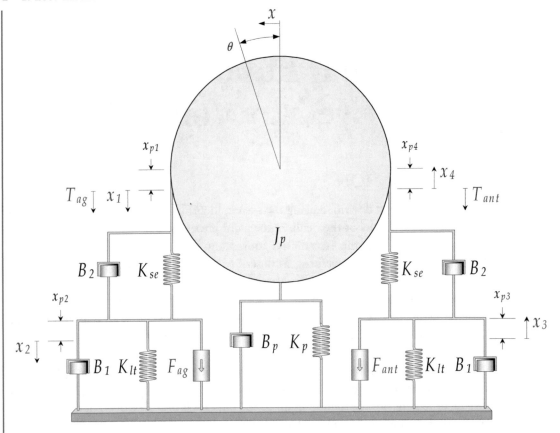

Figure 1.1: Oculomotor plant used for analyzing saccades with post-saccade behavior.

a pulse-slide-step signal during saccadic eye movement (Goldstein, 1983, Optican and Miles, 1985).

It should be noted that the passive elasticity and viscosity of the eyeball in Fig. 1.1 is changed from the model in Section 5.3 of Part 1, which included two Voigt passive elements connected in series to a single Voigt element. The Voigt element with time constant 0.02 s is used in the model presented here. The other Voigt element, with a time constant of 1 s, is neglected, since it has an insignificant effect on the accuracy, as we are modeling a single saccade and not a series of saccades. Further, eliminating this Voigt element reduces the order of the model from 4th- to 3rd-order and simplifies the system identification.

The net torque generated by the muscles during a saccade rotates the eyeball to a new orientation, and, after the saccade is completed, compensates the passive restraining torques generated by orbital tissues.

1.2.1   DERIVATION OF THE DIFFERENTIAL EQUATION DESCRIBING THE OCULOMOTOR SYSTEM

To begin our analysis of this model, we assume:

1. $\dot{x}_2 > \dot{x}_1 = \dot{x} = \dot{x}_4 > \dot{x}_3$

2. elasticity K_p is the passive elasticity for muscles other than the lateral and medical rectus muscle and the eyeball

3. viscous element B_p is due to the friction of the eyeball within the eye socket and for muscles other than the lateral and medical rectus muscle

4. x_i is measured from the equilibrium position

5. zero initial conditions

Note that $x = \theta r$ or $\theta = \frac{x}{r} \times \frac{180}{\pi} = 5.2087 \times 10^3 x$, where x is measured in meters with $r = 11$ mm.

The free body diagrams are shown in Fig. 1.2, which give the node equations for the system as:

Node x: $r B_2(\dot{x}_2 - \dot{x}_1) + r K_{se}(x_2 - x_1) - r B_2(\dot{x}_4 - \dot{x}_3) - r K_{se}(x_4 - x_3) = J_p \ddot{\theta} + B_p \dot{\theta} + K_p \theta,$

$$(1.1)$$

Node 2: $F_{ag} - B_1 \dot{x}_2 - K_{lt} x_2 - B_2(\dot{x}_2 - \dot{x}_1) - K_{se}(x_2 - x_1) = 0,$ $\qquad (1.2)$

Node 3: $F_{ant} + B_1 \dot{x}_3 + K_{lt} x_3 - B_2(\dot{x}_4 - \dot{x}_3) - K_{se}(x_4 - x_3) = 0.$ $\qquad (1.3)$

Note that T_{ag} and T_{ant} are the tensions generated by agonist and antagonist muscles with designated direction as shown in Fig. 1.1, and

$$T_{ag} = B_2(\dot{x}_2 - \dot{x}_1) + K_{se}(x_2 - x_1),$$
$$T_{ant} = B_2(\dot{x}_4 - \dot{x}_3) + K_{se}(x_4 - x_3).$$

$$(1.4)$$

Next, we let

$$J = \frac{J_p}{r} \times 5.2087 \times 10^3, B = \frac{B_p}{r} \times 5.2087 \times 10^3, \text{ and } K = \frac{K_p}{r} \times 5.2087 \times 10^3$$

and rewrite Eq. (1.1) as

$$B_2(\dot{x}_2 + \dot{x}_3 - \dot{x}_1 - \dot{x}_4) + K_{se}(x_2 + x_3 - x_1 - x_4) = J \ddot{x} + B \dot{x} + K x. \qquad (1.5)$$

We have assumed that there is an initial displacement from equilibrium at the primary position for springs K_{lt} and K_{se} since the muscle is 3 mm longer than at equilibrium. That is

$$x_1 = x - x_{p1},$$
$$x_4 = x + x_{p4}.$$

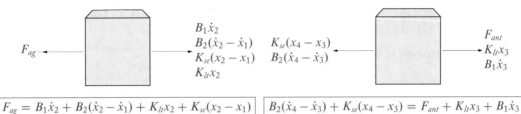

$$F_{ag} = B_1\dot{x}_2 + B_2(\dot{x}_2 - \dot{x}_1) + K_{lt}x_2 + K_{se}(x_2 - x_1)$$

$$B_2(\dot{x}_4 - \dot{x}_3) + K_{se}(x_4 - x_3) = F_{ant} + K_{lt}x_3 + B_1\dot{x}_3$$

$rK_{se}(x_2 - x_1)$
$rB_2(\dot{x}_2 - \dot{x}_1)$

$J_p\ddot{\theta}$
$B_p\dot{\theta}$
$K_p\theta$
$rK_{se}(x_1 - x_3)$
$rB_2(\dot{x}_4 - \dot{x}_3)$

$$rB_2(\dot{x}_2 - \dot{x}_1) + rK_{se}(x_2 - x_1) - rB_2(\dot{x}_4 - \dot{x}_3) - rK_{se}(x_4 - x_3) = J\ddot{\theta} + B\dot{\theta} + K\theta$$

Figure 1.2: Free body diagrams for the system in Fig. 1.1.

To reduce the node equations to a single differential equation, we eliminate variables as before, using the operating point analysis method. To this effect, we introduce the following variables and constants:

$$\hat{x} = x - x(0)$$
$$\hat{\theta} = \theta - \theta(0),$$
$$\hat{x}_1 = x_1 - x_1(0),$$
$$\hat{x}_2 = x_2 - x_2(0),$$
$$\hat{x}_3 = x_3 - x_3(0),$$
$$\hat{x}_4 = x_4 - x_4(0),$$
$$\hat{F}_{ag} = F_{ag} - F_{ag}(0),$$
$$\hat{F}_{ant} = F_{ant} - F_{ant}(0),$$
$$K_{st} = K_{se} + K_{lt},$$
$$B_{12} = B_1 + B_2.$$

Note that with $\hat{x} = \hat{x}_1 = \hat{x}_4$, all derivative terms are zero at steady-state. Eliminating the derivative terms in Eqs. (1.2), (1.3) and (1.5) gives

$$F_{ag}(0) = K_{st}x_2(0) - K_{se}x_1(0),$$
$$F_{ant}(0) = K_{se}x_4(0) - K_{st}x_3(0),$$ (1.6)
$$K_{se}(x_2(0) + x_3(0) - x_1(0) - x_4(0)) = Kx(0).$$

Subtracting the muscle node equations in Eq. (1.6) at steady-state, for use later, gives

$$F_{ag}(0) - F_{ant}(0) = K_{st}(x_2(0) + x_3(0)) - K_{se}(x_1(0) + x_4(0)). \tag{1.7}$$

Substituting the operating point variables and initial conditions into Eq. (1.2) gives

$$\hat{F}_{ag} + F_{ag}(0) - B_1\dot{\hat{x}}_2 - K_{lt}(\hat{x}_2 + x_2(0)) - B_2(\dot{\hat{x}}_2 - \dot{\hat{x}}_1) \\ - K_{se}(\hat{x}_2 + x_2(0) - \hat{x}_1 - x_1(0)) = 0. \tag{1.8}$$

Removing the initial condition terms in Eq. (1.8) using steady-state terms from Eq. (1.6), yields

$$\hat{F}_{ag} = B_1\dot{\hat{x}}_2 + K_{lt}\hat{x}_2 + B_2(\dot{\hat{x}}_2 - \dot{\hat{x}}) + K_{se}(\hat{x}_2 - \hat{x}), \tag{1.9}$$

where $\hat{x}_1$ has been replaced by $\hat{x}$. After repeating this for Eqs. (1.3) and (1.5), we have

$$\hat{F}_{ant} = B_2(\dot{\hat{x}} - \dot{\hat{x}}_3) - B_1\dot{\hat{x}}_3 + K_{se}\hat{x} - K_{st}\hat{x}_3 \quad \text{and} \tag{1.10}$$
$$B_2(\dot{\hat{x}}_2 + \dot{\hat{x}}_3 - 2\dot{\hat{x}}) + K_{se}(\hat{x}_2 + \hat{x}_3 - 2\hat{x}) = J\ddot{\hat{x}} + B\dot{\hat{x}} + K\hat{x}, \tag{1.11}$$

where $\hat{x}_4$ has been replaced by $\hat{x}$. We next apply the Laplace transform to Eqs. (1.9), (1.10) and (1.11), giving

$$\hat{F}_{ag}(s) = \hat{X}_2(s(B_1 + B_2) + K_{st}) - \hat{X}(sB_2 + K_{se}), \tag{1.12}$$
$$\hat{F}_{ant}(s) = \hat{X}(sB_2 + K_{se}) - \hat{X}_3(s(B_1 + B_2) + K_{st}), \tag{1.13}$$
$$(sB_2 + K_{se})(\hat{X}_2 + \hat{X}_3 - 2\hat{X}) = \hat{X}(Js^2 + Bs + K). \tag{1.14}$$

Rearranging Eqs. (1.12) and (1.13) yields

$$\hat{X}_2 = \frac{\hat{F}_{ag}(s) + \hat{X}(sB_2 + K_{se})}{(sB_{12} + K_{st})} \quad \text{and}$$
$$\hat{X}_3 = \frac{(sB_2 + K_{se})\hat{X} - \hat{F}_{ant}(s)}{(sB_{12} + K_{st})},$$

and summing them together yields

$$\hat{X}_2 + \hat{X}_3 = \frac{\hat{F}_{ag}(s) + 2\hat{X}(sB_2 + K_{se}) - \hat{F}_{ant}(s)}{(sB_{12} + K_{st})}. \tag{1.15}$$

Substituting $\hat{X}_2 + \hat{X}_3$ (Eq. (1.15)) into Eq. (1.14) gives

$$(sB_2 + K_{se})\left(\frac{\hat{F}_{ag}(s) + 2\hat{X}(sB_2 + K_{se}) - \hat{F}_{ant}(s)}{(sB_{12} + K_{st})} - 2\hat{X}\right) = \hat{X}(Js^2 + Bs + K). \tag{1.16}$$

After multiplying both sides of Eq. (1.16) by $(sB_{12} + K_{st})$, we have

$$(sB_2 + K_{se})\left(\hat{F}_{ag}(s) - \hat{F}_{ant}(s)\right) + 2\hat{X}(sB_2 + K_{se})^2 \\ - 2\hat{X}(sB_2 + K_{se})(sB_{12} + K_{st}) = \hat{X}(sB_{12} + K_{st})(Js^2 + Bs + K). \tag{1.17}$$

Simplifying Eq. (1.17) gives

$$(sB_2 + K_{se})\left(\hat{F}_{ag}(s) - \hat{F}_{ant}(s)\right) = \hat{X}\left(C_3 s^3 + C_2 s^2 + C_1 s + C_0\right), \tag{1.18}$$

where

$$
\begin{aligned}
C_3 &= JB_{12} \\
C_2 &= JK_{st} + B_{12}B + 2B_1 B_2 \\
C_1 &= 2B_1 K_{se} + 2B_2 K_{lt} + B_{12}K + K_{st}B \\
C_0 &= K_{st}K + 2K_{lt}K_{se}.
\end{aligned}
$$

We now transform back into time domain using the inverse Laplace transform, yielding

$$B_2\left(\dot{\hat{F}}_{ag} - \dot{\hat{F}}_{ant}\right) + K_{se}\left(\hat{F}_{ag} - \hat{F}_{ant}\right) = C_3 \dddot{\hat{x}} + C_2 \ddot{\hat{x}} + C_1 \dot{\hat{x}} + C_0 \hat{x}. \tag{1.19}$$

With $\dot{F}_{ag} = \dot{\hat{F}}_{ag}$, $\dot{F}_{ant} = \dot{\hat{F}}_{ant}$, $\hat{F}_{ag} = F_{ag} - F_{ag}(0)$ and $\hat{F}_{ant} = F_{ant} - F_{ant}(0)$, we have

$$\hat{F}_{ag} - \hat{F}_{ant} = F_{ag} - F_{ag}(0) - F_{ant} + F_{ant}(0). \tag{1.20}$$

From Eq. (1.7), we have

$$F_{ag}(0) - F_{ant}(0) = K_{st}\left(x_2(0) + x_3(0)\right) - K_{se}\left(x_1(0) + x_4(0)\right)$$

and with

$$x_1(0) = x(0) - x_{p1} \text{ and } x_4(0) = x(0) + x_{p4} = x(0) + x_{p1}$$

and by assuming identical muscles, we have

$$F_{ag}(0) - F_{ant}(0) = K_{st}\left(x_2(0) + x_3(0)\right) - 2K_{se}x(0). \tag{1.21}$$

From Eq. (1.6), we have

$$K_{se}\left(x_2(0) + x_3(0) - x_1(0) - x_4(0)\right) = Kx(0)$$

and after removing $x_1(0)$ and $x_4(0)$ as before, this gives

$$K_{se}\left(x_2(0) + x_3(0) - 2x(0)\right) = Kx(0)$$

and rearranging the previous equation, yields

$$x_2(0) + x_3(0) = \left(\frac{K}{K_{se}} + 2\right)x(0).$$

When $x_2(0) + x_3(0)$ is substituted into Eq. (1.21), we have

$$F_{ag}(0) - F_{ant}(0) = \left(K_{st}\left(\frac{K}{K_{se}} + 2\right) - 2K_{se}\right)x(0). \tag{1.22}$$

With Eqs. (1.20) and (1.22) inserted into Eq. (1.19), we have

$$B_2 \left(\dot{F}_{ag} - \dot{F}_{ant} \right) + K_{se} \left(F_{ag} - F_{ant} \right) - K_{se} \left(F_{ag}(0) - F_{ant}(0) \right)$$
$$= C_3 \dddot{x} + C_2 \ddot{x} + C_1 \dot{x} + C_0 \hat{x}.$$

Then with Eq. (1.22) $\left(F_{ag}(0) - F_{ant}(0) \right)$ substituted into the previous equation, we have

$$B_2 \left(\dot{F}_{ag} - \dot{F}_{ant} \right) + K_{se} \left(F_{ag} - F_{ant} \right) - K_{se} \left(K_{st} \left(\frac{K}{K_{se}} + 2 \right) - 2 K_{se} \right) x(0).$$
$$= C_3 \dddot{x} + C_2 \ddot{x} + C_1 \dot{x} + C_0 \left(\hat{x} - x(0) \right). \tag{1.23}$$

To reduce Eq. (1.23) further, we note that

$$K_{se} \left(K_{st} \left(\frac{K}{K_{se}} + 2 \right) - 2 K_{se} \right) x(0) = C_0 x(0) = (K_{st} K + 2 K_{lt} K_{se}) x(0),$$

or

$$K_{st} K + 2 K_{st} - 2 K_{se} = K_{st} K + 2 K_{lt} K_{se}. \tag{1.24}$$

Since the left-hand side of Eq. (1.24) equals the right-hand side of Eq. (1.24), Eq. (1.23) becomes

$$B_2 \left(\dot{F}_{ag} - \dot{F}_{ant} \right) + K_{se} \left(F_{ag} - F_{ant} \right) = C_3 \dddot{x} + C_2 \ddot{x} + C_1 \dot{x} + C_0 \hat{x}. \tag{1.25}$$

In Eq. (1.25), we exchange x for θ, $\left(x = \frac{\theta}{5208.7} \right)$, giving

$$\delta \left(B_2 \left(\dot{F}_{ag} - \dot{F}_{ant} \right) + K_{se} \left(F_{ag} - F_{ant} \right) \right) = \dddot{\theta} + P_2 \ddot{\theta} + P_1 \dot{\theta} + P_0 \theta, \tag{1.26}$$

where

$$\delta = \frac{5208.7}{J B_{12}}$$
$$P_2 = \frac{J K_{st} + B_{12} B + 2 B_1 B_2}{J B_{12}}$$
$$P_1 = \frac{2 B_1 K_{se} + 2 B_2 K_{lt} + B_{12} K + K_{st} B}{J B_{12}}$$
$$P_0 = \frac{K_{st} K + 2 K_{lt} K_{se}}{J B_{12}}.$$

A block diagram for the system in Fig. 1.1 is shown in Fig. 1.3. The section of the diagram in the forward path is the 2009 linear homeomorphic saccadic eye movement model. The feedback element H is unity and is operational only when the eye is stable. It is clear that this model has a characteristic equation that is 3rd-order rather than 4th-order in the previous version of the model. This model is easier to use than the 1995 model without sacrificing accuracy.

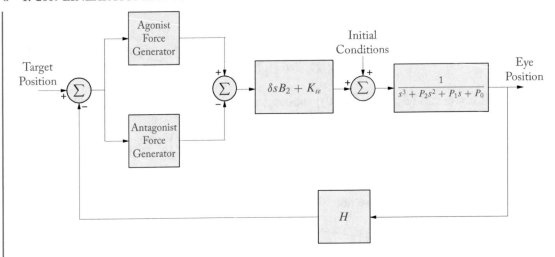

Figure 1.3: Block diagram of the 2009 linear homeomorphic saccadic eye movement model.

1.2.2   NEURAL INPUT

In Section 3.3 of Part 1, we modeled the neural input to the saccade system as a pulse-step waveform. This input has been used in many studies because of its simplicity and ease of use (Bahill et al., 1980, Enderle et al., 1984, Enderle and Wolfe, 1988a). To create a more realistic input based on physiological evidence, a pulse-slide-step input is used, as shown in Fig. 1.4 (based on Goldstein (1983)). The slide is an exponential transition from the pulse to the step. This model is consistent with the data published in the literature (for example, see Fig. 4 in Robinson (1981) and Fig. 2 in Van Gisbergen et al. (1981)). The diagram in Fig. 1.4 (top) closely approximates the data shown in Fig. 1.4 (bottom) for the agonist input. An explanation of the neural input will be given in a later section.

At steady-state, the eye is held steady by the agonist and antagonist inputs F_{g0} and F_{t0}. We typically define the time when the target moves as $t = 0$. This is a common assumption since many simulation studies ignore the latent period and focus on the actual movement (see the time axis in Fig. 1.4, top and bottom).

The overall agonist pulse occurs in the interval $0 - T_2$, with a more complex behavior than the pulse described in Section 3.3 of Part 1. We view the overall pulse process as the intention of the system, which is limited by its physical capabilities. The start of the pulse occurs with an exponential rise from the initial firing rate, F_{g0}, to peak magnitude, F_{p1}, with a time constant τ_{gn1}. At T_1, the input decays to F_{p2}, with a time constant τ_{gn2}. The slide occurs at T_2, with a time constant τ_{gn3}, to F_{gs}, the force necessary to hold the eye at its destination. The input F_{gs} is applied during the step portion of the input.

At $t = 0$, the antagonist neural input is completely inhibited and exponentially decays to zero from F_{t0}, with time constant τ_{tn1}. At time T_3, the antagonist input exponentially increases

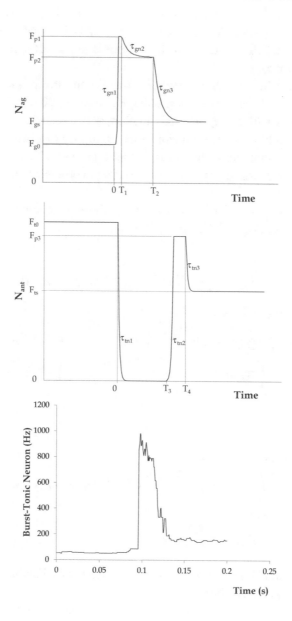

Figure 1.4: Neural input to the horizontal saccade system. (top) Agonist input. (middle) Antagonist input. (bottom) Discharge rate of a single burst-tonic neuron during a saccade (agonist input). Details of the experiment and training for (bottom) are reported elsewhere (Sparks et al., 1976; data provided personally by Dr. David Sparks).

with time constant τ_{tn2}. The antagonist neural input shown in Fig. 1.4 (middle) includes a PIRB pulse with duration of $T_4 - T_3$. At T_4, the antagonist input exponentially decays to F_{ts}, with a time constant τ_{tn3}. If no PIRB occurs in the antagonist input, the input exponentially rises to F_{ts}, with time constant τ_{tn2}.

The agonist pulse includes an interval (T_1) that is constant for saccades of all sizes as supported by physiological evidence (Enderle, 2002, Zhou et al., 2009)). We choose to model the change in the firing rate with an exponential function, as this seems to match the data fairly well.

After complete inhibition, the antagonist neural input has a brief excitatory pulse starting at T_3 with duration of approximately 10 ms. Enderle proposed that this burst is generated by PIRB, a property which contributes to the post-saccade phenomena such as dynamic and glissadic overshoot (2002).

Based on the diagram in Fig. 1.4 and assuming the exponential terms reach steady-state at $t = 5\tau$, the equations for N_{ag} and N_{ant} are written as:

$$
\begin{aligned}
N_{ag} = {} & F_{g0}u(-t) \\
& + \left(F_{g0} + (F_{p1} - F_{g0})e^{\frac{t}{\tau_{gn1}} - 5} \right) \left(u(t) - u\left(t - 5\tau_{gn1} \right) \right) \\
& + F_{p1} \left(u\left(t - 5\tau_{gn1} \right) - u(t - T_1) \right) \\
& + \left(F_{p2} + (F_{p1} - F_{p2})e^{\frac{(T_1 - t)}{\tau_{gn2}}} \right) \left(u(t - T_1) - u(t - T_2) \right) \\
& + \left(F_{gs} + \left(F_{p2} + (F_{p1} - F_{p2})\, e^{\frac{(T_1 - T_2)}{\tau_{gn2}}} - F_{gs} \right) e^{\frac{(T_2 - t)}{\tau_{gn3}}} \right) u(t - T_2)
\end{aligned}
\tag{1.27}
$$

and

$$
\begin{aligned}
N_{ant} = {} & F_{t0}u(-t) \\
& + F_{t0}e^{\frac{-t}{\tau_{tn1}}} \left(u(t) - u(t - T_3) \right) \\
& + \left(F_{t0}e^{\frac{-T_3}{\tau_{tn1}}} + \left(F_{p3} - F_{t0}e^{\frac{-T_3}{\tau_{tn1}}} \right) e^{\frac{(t - T_3)}{\tau_{tn2}} - 5} \right) \left(u(t - T_3) - u(t - T_3 - 5\tau_{tn2}) \right) \\
& + F_{p3} \left(u(t - T_3 - 5\tau_{tn2}) - u(t - T_4) \right) \\
& + \left(F_{ts} + (F_{p3} - F_{ts})\, e^{\frac{(T_4 - t)}{\tau_{tn3}}} \right) u(t - T_4).
\end{aligned}
\tag{1.28}
$$

Note that Eqs. (1.27) and (1.28) are written in terms of intervals. Further, we assume that $5\tau_{gn1} < T_1$ and $T_3 + 5\tau_{tn2} < T_4$, which simplifies analysis.

As before, the agonist and antagonist active-state tensions are defined as low-pass filtered neural inputs:

$$
\dot{F}_{ag} = \frac{N_{ag} - F_{ag}}{\tau_{ag}}
\tag{1.29}
$$

$$
\dot{F}_{ant} = \frac{N_{ant} - F_{ant}}{\tau_{ant}}
\tag{1.30}
$$

where

$$\tau_{ag} = \tau_{gac}\left(u(t) - u\left(t - T_2\right)\right) + \tau_{gde}u\left(t - T_2\right) \tag{1.31}$$

$$\tau_{ant} = \tau_{tde}\left(u(t) - u\left(t - T_3\right)\right) + \tau_{tac}\left(u\left(t - T_3\right) - u\left(t - T_4\right)\right) + \tau_{tde}u\left(t - T_4\right) \tag{1.32}$$

The activation and deactivation time constants represent the different dynamic characteristics of muscle under increasing and decreasing stimulation. Shown in Fig. 1.5 are the agonist and antagonist active-state tensions, derived from low-pass filtering the neural inputs.

The analytical solutions for F_{ag} and F_{ant} are derived from Eqs. (1.29) and (1.30) using the neural inputs from Eqs. (1.27) and (1.28). We use following constants for ease in solution:

$$A = F_{p1} - F_{g0}$$
$$B = F_{p1} - F_{p2}$$
$$C = F_{p2} + \left(F_{p1} - F_{p2}\right)e^{\frac{(T_1 - T_2)}{\tau_{gn2}}} - F_{gs}$$
$$D = F_{t0}e^{\frac{-T_3}{\tau_{tn1}}}$$
$$E = F_{p3} - F_{t0}e^{\frac{-T_3}{\tau_{tn1}}}$$
$$F = F_{p3} - F_{ts}.$$

F_{ag} Response   The agonist *natural response* has one time constant, τ_{ag}, and, in general, has a response of $F_{ag_n} = K_g e^{-\frac{t}{\tau_{ag}}}$. Since τ_{ag} is defined in two intervals according to Eq. (1.31), we need to be careful in how this is handled. There are four inputs applied to the system, so we use superposition to write the solution.

$$
\begin{aligned}
F_{ag_n}(t) = {}& K_{g0}e^{\frac{-t}{\tau_{gac}}}u(t) - u(t) - 5\tau_{gn1} \\
&+ K_{g1}e^{\frac{5\tau_{gn1}-t}{\tau_{gac}}}u(t) - 5\tau_{gn1} - u(t) - T_1 \\
&+ K_{g2}e^{\frac{T_1-t}{\tau_{gac}}}u(t) - T_1 - u(t) - T_2 \\
&+ K_{g3}e^{\frac{T_2-t}{\tau_{gde}}}u(t - T_2),
\end{aligned}
\tag{1.33}
$$

where K_{g0}, K_{g1}, K_{g2} and K_{g3} are unknown constants to be determined from the initial conditions.

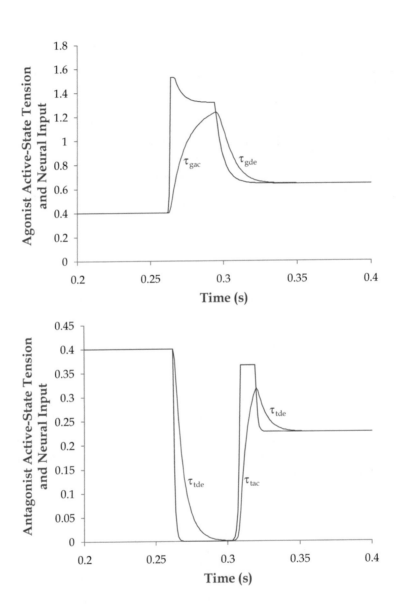

Figure 1.5: (Top) Agonist active-state tension (red) and neural input (blue). (Bottom) Antagonist active-state tension (red) and neural input (blue).

The agonist *forced response* is based on the form of the input. We substitute the input for each interval in Eq. (1.27) into Eq. (1.29), and find, after some work, that

$$
\begin{aligned}
F_{ag_f}(t) = \; & F_{g0}u(-t) \\
& + \left(F_{g0} + L_{g1}e^{\frac{t}{\tau_{gn1}} - 5} \right)\left(u(t) - u\left(t - 5\tau_{gn1} \right) \right) \\
& + F_{p1}\left(u\left(t - 5\tau_{gn1} \right) - u\left(t - T_1 \right) \right) \\
& + \left(F_{p2} + L_{g2}e^{\frac{(T_1 - t)}{\tau_{gn2}}} \right)\left(u\left(t - T_1 \right) - u\left(t - T_2 \right) \right) \\
& + \left(F_{gs} + L_{g3}e^{\frac{(T_2 - t)}{\tau_{gn3}}} \right) u\left(t - T_2 \right),
\end{aligned}
\tag{1.34}
$$

where

$$
\begin{aligned}
L_{g1} &= \frac{\tau_{gn1}}{\tau_{gac} + \tau_{gn1}} A \\
L_{g2} &= \frac{\tau_{gn2}}{\tau_{gn2} - \tau_{gac}} B \\
L_{g3} &= \frac{\tau_{gn3}}{\tau_{gn3} - \tau_{gde}} C.
\end{aligned}
$$

The total agonist active-state tension equals $F_{ag} = F_{ag_n} + F_{ag_f}$ and is given by

$$
\begin{aligned}
F_{ag}(t) = \; & K_{g0}e^{\frac{-t}{\tau_{gac}}}\left(u(t) - u\left(t - 5\tau_{gn1} \right) \right) + K_{g1}e^{\frac{(5\tau_{gn1} - t)}{\tau_{gac}}}\left(u\left(t - 5\tau_{gn1} \right) - u\left(t - T_1 \right) \right) \\
& + K_{g2}e^{\frac{(T_1 - t)}{\tau_{gac}}}\left(u\left(t - T_1 \right) - u\left(t - T_2 \right) \right) + K_{g3}e^{\frac{(T_2 - t)}{\tau_{gde}}} u(t - T_2) + F_{g0}u(-t) \\
& + \left(F_{g0} + L_{g1}e^{\frac{t}{\tau_{gn1}} - 5} \right)\left(u(t) - u\left(t - 5\tau_{gn1} \right) \right) + F_{p1}\left(u(t - 5\tau_{gn1}) - u\left(t - T_1 \right) \right) \\
& + \left(F_{p2} + L_{g2}e^{\frac{(T_1 - t)}{\tau_{gn2}}} \right)\left(u\left(t - T_1 \right) - u\left(t - T_2 \right) \right) + \left(F_{gs} + L_{g3}e^{\frac{(T_2 - t)}{\tau_{gn3}}} \right) u\left(t - T_2 \right) \\
= \; & F_{g0}u(-t) + \left(K_{g0}e^{\frac{-t}{\tau_{gac}}} + F_{g0} + L_{g1}e^{\frac{t}{\tau_{gn1}} - 5} \right)\left(u(t) - u\left(t - 5\tau_{gn1} \right) \right) \\
& + \left(K_{g1}e^{\frac{(5\tau_{gn1} - t)}{\tau_{gac}}} + F_{p1} \right)\left(u\left(t - 5\tau_{gn1} \right) - u\left(t - T_1 \right) \right) \\
& + \left(K_{g2}e^{\frac{(T_1 - t)}{\tau_{gac}}} + F_{p2} + L_{g2}e^{\frac{(T_1 - t)}{\tau_{gn2}}} \right)\left(u\left(t - T_1 \right) - u\left(t - T_2 \right) \right) \\
& + \left(K_{g3}e^{\frac{(T_2 - t)}{\tau_{gde}}} + F_{gs} + L_{g3}e^{\frac{(T_2 - t)}{\tau_{gn3}}} \right) u\left(t - T_2 \right).
\end{aligned}
\tag{1.35}
$$

To evaluate unknown constants, K_{g0}, K_{g1}, K_{g2} and K_{g3}, in Eq. (1.35), we use the initial condition and the fact that F_{ag} is continuous in time. For example, at switch time t_1, $F_{ag}\left(T_1^- \right) =$

$F_{ag}\left(T_1^+\right)$. Thus,

$$t = 0: \quad F_{ag} = F_{g0}$$
$$F_{g0} = K_{g0} + F_{g0} + L_{g1}e^{-5} \Rightarrow K_{g0} = F_{g0} - F_{g0} + L_{g1}e^{-5}$$

$$t = 5\tau_{gn1}: \quad F_{ag} \text{ is continuous}$$
$$K_{g0}e^{-\frac{5\tau_{gn1}}{\tau_{gac}}} + F_{g0} + L_{g1} = K_{g1} + F_{p1} \Rightarrow K_{g1} = K_{g0}e^{-\frac{5\tau_{gn1}}{\tau_{gac}}} + F_{g0} + L_{g1} - F_{p1}$$

$$t = T_1: \quad F_{ag} \text{ is continuous}$$
$$K_{g1}e^{-\frac{T_1 - 5\tau_{gn1}}{\tau_{gac}}} + F_{p1} = K_{g2} + F_{p2} + L_{g2} \Rightarrow K_{g2} = K_{g1}e^{-\frac{T_1 - 5\tau_{gn1}}{\tau_{gac}}} + F_{p1} - F_{p2} + L_{g2}$$

$$t = T_2: \quad F_{ag} \text{ is continuous}$$
$$K_{g2}e^{\frac{T_1 - T_2}{\tau_{gac}}} + F_{p2} + L_{g2}e^{\frac{T_1 - T_2}{\tau_{gn2}}} = K_{g3} + F_{gs} + L_{g3}$$
$$\Rightarrow K_{g3} = K_{g2}e^{\frac{T_1 - T_2}{\tau_{gac}}} + F_{p2} + L_{g2}e^{\frac{T_1 - T_2}{\tau_{gn2}}} - F_{gs} + L_{g3}.$$

Based on these calculations and simplifying Eq. (1.35) by bringing together the unit step functions as single terms, F_{ag} is

$$
\begin{aligned}
F_{ag} = {} & F_{g0}u(-t) \\
& + \left(K_{g0}e^{\frac{-t}{\tau_{gac}}} + F_{g0} + L_{g1}e^{\frac{t}{\tau_{gn1}}-5}\right)u(t) \\
& + \left(M_{g1}e^{\frac{(5\tau_{gn1}-t)}{\tau_{gac}}} + F_{p1} - F_{g0} - L_{g1}e^{\frac{t}{\tau_{gn1}}-5}\right)u\left(t - 5\tau_{gn1}\right) \qquad (1.36) \\
& + \left(M_{g2}e^{\frac{(T_1-t)}{\tau_{gac}}} + F_{p2} + L_{g2}e^{\frac{(T_1-t)}{\tau_{gn2}}} - F_{p1}\right)u\left(t - T_1\right) \\
& + \left(M_{g3}e^{\frac{(T_2-t)}{\tau_{gac}}} + K_{g3}e^{\frac{(T_2-t)}{\tau_{gde}}} + F_{gs} + L_{g3}e^{\frac{(T_2-t)}{\tau_{gn3}}} - F_{p2} - L_{g2}e^{\frac{(T_1-t)}{\tau_{gn2}}}\right)u\left(t - T_2\right),
\end{aligned}
$$

where

$$M_{g1} = K_{g1} - K_{g0}e^{-\frac{5\tau_{gn1}}{\tau_{gac}}}$$

$$M_{g2} = K_{g2} - K_{g1}e^{\frac{5\tau_{gn1}-T_1}{\tau_{gac}}}$$

$$M_{g3} = - K_{g2}e^{\frac{T_1-T_2}{\tau_{gac}}}.$$

F_{ant} Response   The antagonist *natural response* has one time constant, τ_{ant} that is defined in two intervals as in Eq. (1.32). As before, with several inputs applied, we use superposition to write the solution.

$$F_{ant_n} = K_{t0}e^{-\frac{t}{\tau_{tde}}} \left(u(t) - u(t - T_3)\right)$$

$$+ K_{t1}e^{-\frac{t-T_3}{\tau_{tac}}} \left(u(t - T_3) - u(t - T_3 - 5\tau_{tn2})\right)$$

$$+ K_{t2}e^{-\frac{t-T_3-5\tau_{tn2}}{\tau_{tac}}} \left(u(t - T_3 - 5\tau_{tn2}) - u(t - T_4)\right) \tag{1.37}$$

$$+ K_{t3}e^{-\frac{t-T_4}{\tau_{tde}}} u(t - T_4),$$

where K_{t0}, K_{t1}, K_{t2} and K_{t3} are unknown constants to be determined from the initial conditions.

The antagonist *forced response* is based on the form of the input. We substitute the input for each interval in Eq. (1.28) into Eq. (1.30), and have

$$F_{ant_f}(t) = F_{t0}u(-t)$$

$$+ L_{t1}e^{\frac{-t}{\tau_{m1}}} \left(u(t) - u(t - T_3)\right)$$

$$+ \left(D + L_{t2}e^{\frac{t-T_3}{\tau_{m2}}-5}\right) \left(u(t - T_3) - u(t - T_3 - 5\tau_{tn2})\right) \tag{1.38}$$

$$+ F_{p3} \left(u(t - T_3 - 5\tau_{tn2}) - u(t - T_4)\right)$$

$$+ \left(F_{ts} + L_{t3}e^{\frac{T_4-t}{\tau_{tn3}}}\right) u(t - T_4),$$

where

$$L_{t1} = \frac{\tau_{tn1}}{\tau_{tn1} - \tau_{tde}} F_{t0}$$

$$= \frac{\tau_{tn2}}{\tau_{tn2} - \tau_{tac}} E$$

$$= \frac{\tau_{tn3}}{\tau_{tn3} - \tau_{tde}} F.$$

The total antagonist active-state tension equals $F_{ant} = F_{ant_n} + F_{ant_f}$ and is given by

$$F_{ant}(t) = K_{t0}e^{\frac{-t}{\tau_{tde}}}\left(u(t) - u(t - T_3)\right) + K_{t1}e^{\frac{(T_3-t)}{\tau_{tac}}}\left(u(t - T_3) - u(t - T_3 - 5\tau_{tn2})\right)$$
$$+ K_{t2}e^{\frac{T_3 + 5\tau_{tn2} - t}{\tau_{tac}}}\left(u(t - T_3 - 5\tau_{tn2}) - u(t - T_4)\right) + K_{t3}e^{\frac{(T_4-t)}{\tau_{tde}}}u(t - T_4) + F_{t0}u(-t)$$
$$+ L_{t1}e^{\frac{-t}{\tau_{tn1}}}\left(u(t) - u(t - T_3)\right) + \left(D + L_{t2}e^{\frac{t - T_3}{\tau_{tn2}} - 5}\right)\left(u(t - T_3) - u(t - T_3 - 5\tau_{tn2})\right)$$
$$+ F_{p3}\left(u(t - T_3 - 5\tau_{tn2}) - u(t - T_4)\right) + \left(F_{ts} + L_{t3}e^{\frac{(T_4-t)}{\tau_{tn3}}}\right)u(t - T_4)$$

$$= F_{t0}u(-t) + \left(K_{t0}e^{\frac{-t}{\tau_{tde}}} + L_{t1}e^{\frac{-t}{\tau_{tn1}}}\right)(u(t) - u(t - T_3))$$
$$+ \left(K_{t1}e^{\frac{(T_3-t)}{\tau_{tac}}} + D + L_{t2}e^{\frac{t - T_3}{\tau_{tn2}} - 5}\right)\left(u(t - T_3) - u(t - T_3 - 5\tau_{tn2})\right) \qquad (1.39)$$
$$+ \left(K_{t2}e^{\frac{(T_3 + 5\tau_{tn2} - t)}{\tau_{tac}}} + F_{p3}\right)\left(u(t - T_3 - 5\tau_{tn2}) - u(t - T_4)\right)$$
$$+ \left(K_{t3}e^{\frac{(T_4-t)}{\tau_{tde}}} + F_{ts} + L_{t3}e^{\frac{(T_4-t)}{\tau_{tn3}}}\right)u(t - T_4).$$

To evaluate unknown constants, K_{t0}, K_{t1}, K_{t2}, and K_{t3}, in Eq. (1.39), we use the initial condition and the fact that F_{ant} is continuous in time. For example, at switch time T_3, $F_{ant}\left(T_3^-\right) = F_{ant}\left(T_3^+\right)$. Thus,

$t = 0 : F_{ant} = F_{t0}$
$$F_{t0} = K_{t0} + L_{t1} \Rightarrow K_{t0} = F_{t0} - L_{t1}$$

$t = T_3 : F_{ant}$ is continuous
$$L_{t1}e^{\frac{-T_3}{\tau_{tn1}}} + K_{t0}e^{\frac{T_3}{\tau_{tde}}} = K_{t1} + D + L_{t2}e^{-5} \Rightarrow K_{t1} = L_{t1}e^{\frac{T_3}{\tau_{tn1}}} + K_{t0}e^{\frac{T_3}{\tau_{tde}}} - \left(D + L_{t2}e^{-5}\right)$$

$t = T_3 + 5\tau_{tn2} : F_{ant}$ is continuous
$$K_{t1}e^{\frac{-5\tau_{tn2}}{\tau_{tac}}} + D + L_{t2} = K_{t2} + F_{p3} \Rightarrow K_{t2} = K_{t1}e^{\frac{-5\tau_{tn2}}{\tau_{tac}}} + D + L_{t2} - F_{p3}$$

$t = T_4 : F_{ant}$ is continuous
$$K_{t2}e^{\frac{(T_3 + 5\tau_{tn2} - T_4)}{\tau_{tac}}} + F_{p3} = K_{t3} + F_{ts} + L_{t3}$$
$$\Rightarrow K_{t3} = K_{t2}e^{\frac{(T_3 + 5\tau_{tn2} - T_4)}{\tau_{tac}}} + F_{p3} - (F_{ts} + L_{t3}).$$

Based on these calculations and simplifying Eq. (1.38) by bringing together the unit step functions as single terms, F_{ant} is

$$
\begin{aligned}
F_{ant}(t) = {}& F_{t0}u(-t) \\
& + \left(K_{t0}e^{-\frac{t}{\tau_{tde}}} + L_{t1}e^{-\frac{t}{\tau_{tn1}}} \right) u(t) \\
& + \left(K_{t1}e^{\frac{(T_3-t)}{\tau_{tac}}} - M_{t1}e^{\frac{(T_3-t)}{\tau_{tde}}} + D + L_{t2}e^{\frac{(t-T_3)}{\tau_{tn2}}-5} - L_{t1}e^{\frac{-t}{\tau_{tn1}}} \right) u(t-T_3) \\
& + \left(M_{t2}e^{\frac{(T_3+5\tau_{tn2}-t)}{\tau_{tac}}} + F_{p3} - D - L_{t2}e^{\frac{(t-T_3)}{\tau_{tn2}}-5} \right) u(t-T_3-5\tau_{tn2}) \\
& + \left(M_{t3}e^{\frac{(T_4-t)}{\tau_{tac}}} + K_{t3}e^{\frac{(T_4-t)}{\tau_{tde}}} - F + L_{t3}e^{\frac{(T_4-t)}{\tau_{tn3}}} \right) u(t-T_4),
\end{aligned}
\tag{1.40}
$$

where

$$
\begin{aligned}
M_{t1} &= K_{t0}e^{\frac{-T_3}{\tau_{tde}}} \\
M_{t2} &= K_{t2} - K_{t1}e^{\frac{-5\tau_{tn2}}{\tau_{tac}}} \\
M_{t3} &= - K_{t2}e^{\frac{(T_3+5\tau_{tn2}-T_4)}{\tau_{tac}}}.
\end{aligned}
$$

The derivatives of active-state tensions are found by using Eqs. (1.29) and (1.30). F_{g0}, F_{gs}, F_{t0} and F_{ts} are steady-state tensions determined from:

$$
F = \begin{cases} 0.4 + 0.0175\,|\theta|\ N & \text{for } \theta \geq 0° \\ 0.4 - 0.0125\,|\theta|\ N & \text{for } \theta < 0° \end{cases}.
$$

In practice, a short time delay of 1 ms is introduced to reflect the time it takes to send the signal from the abducens and oculomotor nuclei to the muscles.

1.2.3   SACCADE RESPONSE

At this time, we wish to solve for the complete response for a saccade. To begin, the system given by Eq. (1.26) is repeated here for convenience as

$$
\delta \left(B_2 \left(\dot{F}_{ag} - \dot{F}_{ant} \right) + K_{se} \left(F_{ag} - F_{ant} \right) \right) = \dddot{\theta} + P_2\ddot{\theta} + P_1\dot{\theta} + P_0\theta,
$$

which is driven by the inputs F_{ag} and F_{ant} described by Eqs. (1.36) and (1.40). As usual, the solution is composed of the natural and forced response. We begin with the forced response and follow with the natural response.

Forced Response   To evaluate the forced response for Eq. (1.26), Eqs. (1.35), (1.39) and derivative terms are substituted into the left-hand side of Eq. (1.27). The forced response is a linear combination of unit step functions with exponential components. The unit step functions are $u(t)$,

$u\left(t - 5\tau_{gn1}\right)$, $u\left(t - T_1\right)$, $u\left(t - T_2\right)$, $u\left(t - T_3\right)$, $u\left(t - T_3 - 5\tau_{tn2}\right)$, and $u\left(t - T_4\right)$. The exponential components are $e^{\frac{(t_n - t)}{\tau_{gac}}}$, $e^{\frac{(t_n - t)}{\tau_{gde}}}$, $e^{\frac{(t_n - t)}{\tau_{gn1}}}$, $e^{\frac{(t_n - t)}{\tau_{gn2}}}$, $e^{\frac{(t_n - t)}{\tau_{gn3}}}$, $e^{\frac{(t_n - t)}{\tau_{tac}}}$, $e^{\frac{(t_n - t)}{\tau_{tde}}}$, $e^{\frac{(t_n - t)}{\tau_{tn1}}}$, $e^{\frac{(t_n - t)}{\tau_{tn2}}}$, and $e^{\frac{(t_n - t)}{\tau_{tn3}}}$; the variable t_n represents the switch times. For example, we can write the left-hand side of Eq. (1.26) as

$$
LHS = \delta \begin{bmatrix} \left(C_{1,1} + C_{1,2}e^{\frac{-t}{\tau_{gac}}} + C_{1,3}e^{\frac{-t}{\tau_{gde}}} + \ldots + C_{1,11}e^{\frac{-t}{\tau_{tn3}}}\right)u(t) + \\ \left(C_{2,1} + C_{2,2}e^{\frac{(5\tau_{gn1} - t)}{\tau_{gac}}} + C_{2,3}e^{\frac{(5\tau_{gn1} - t)}{\tau_{gde}}} + \ldots + C_{2,11}e^{\frac{(5\tau_{gn1} - t)}{\tau_{tn3}}}\right)u\left(t - 5\tau_{gn1}\right) + \\ \vdots \\ \left(C_{7,1} + C_{7,2}e^{\frac{(T_4 - t)}{\tau_{gac}}} + C_{7,3}e^{\frac{(T_4 - t)}{\tau_{gde}}} + \ldots + C_{7,11}e^{\frac{(T_4 - t)}{\tau_{tn3}}}\right)u\left(t - T_4\right) \end{bmatrix}
$$

$$(1.41)$$

where $C_{m,n}$ are the coefficients for each term whose values are listed in Table 1.1.

The forced response is

$$
\theta_f(t) = \delta \begin{pmatrix} \left(A_{1,1} + A_{1,2}e^{-t/\tau_{gac}} + A_{1,3}e^{-t/\tau_{gde}} + \ldots + A_{1,11}e^{-t/\tau_{tn3}}\right)u(t) \\[2mm] + \begin{pmatrix} A_{2,1} + A_{2,2}e^{-(t-5\tau_{gn1})/\tau_{gac}} + A_{2,3}e^{-(t-5\tau_{gn1})/\tau_{gde}} + \ldots \\ + A_{2,11}e^{-(t-5\tau_{gn1})/\tau_{tn3}} \end{pmatrix} u\left(t - 5\tau_{gn1}\right) \\[2mm] \vdots \\[2mm] + \begin{pmatrix} A_{7,1} + A_{7,2}e^{-(t-T_4)/\tau_{gac}} + A_{7,3}e^{-(t-T_4)/\tau_{gde}} + \ldots \\ + A_{7,11}e^{-(t-T_4)/\tau_{tn3}} \end{pmatrix} u\left(t - T_4\right) \end{pmatrix}
$$

$$(1.42)$$

where the coefficients $A_{m,n}$ are functions of $C_{m,n}$ as shown in Table 1.1 by considering the right-hand side of Eq. (1.26). For example, $A_{1,n(n=1\ldots11)} = \frac{\delta C_{1,n}}{D_n}$, where D_n is

$$D_1 = R_0;$$

$$D_2 = D_{gac} = -\frac{1}{\tau_{gac}^3} + \frac{R_2}{\tau_{gac}^2} - \frac{R_1}{\tau_{gac}} + R_0;$$

$$D_3 = D_{gde} = -\frac{1}{\tau_{gde}^3} + \frac{R_2}{\tau_{gde}^2} - \frac{R_1}{\tau_{gde}} + R_0;$$

Table 1.1: Coefficients of the terms in Eq. (1.41) (i.e., $C_{1,1} = K_{se}F_{g0}$, $C_{1,2} = \left(K_{se} - \frac{B_2}{\tau_{gac}}\right)K_{g0}$, $C_{2,1} = K_{se}(F_{p1} - F_{g0})$, etc.)

LHS $= \delta\Sigma$	1	$\exp\left(\frac{t_u - t}{\tau_{gac}}\right)$	$\exp\left(\frac{t_u - t}{\tau_{gde}}\right)$	$\exp\left(\frac{t_u - t}{\tau_{gn1}}\right)$	$\exp\left(\frac{t_u - t}{\tau_{gn2}}\right)$	$\exp\left(\frac{t_u - t}{\tau_{gn3}}\right)$
$u(t)$	$K_{se}F_{g0}$	$\left(K_{se} - \frac{B_2}{\tau_{gac}}\right)K_{g0}$	0	$\left(K_{se} + \frac{B_2}{\tau_{gn1}}\right)L_{g1}e^{-5}$	0	0
$ut - 5\tau_{gn1}$	$K_{se}F_{p1} - F_{g0}$	$\left(K_{se} - \frac{B_2}{\tau_{gac}}\right)M_{g1}$	0	$-\left(K_{se} + \frac{B_2}{\tau_{gn1}}\right)L_{g1}$	0	0
$ut - T_1$	$-K_{se}B$	$\left(K_{se} - \frac{B_2}{\tau_{gac}}\right)M_{g2}$	0	0	$\left(K_{se} - \frac{B_2}{\tau_{gn2}}\right)L_{g2}$	0
$ut - T_2$	$K_{se}F_{g3} - F_{p2}$	$\left(K_{se} - \frac{B_2}{\tau_{gac}}\right)M_{g3}$	$\left(K_{se} - \frac{B_2}{\tau_{gde}}\right)K_{g3}$	0	$-\left(K_{se} - \frac{B_2}{\tau_{gn2}}\right)L_{g2}e^{-\frac{t_2-t_1}{\tau_{gn2}}}$	$\left(K_{se} - \frac{B_2}{\tau_{gn3}}\right)L_{g3}$
$ut - T_3$	$-K_{se}D$	0	0	0	0	0
$ut - T_3 - 5\tau_{tn2}$	$-K_{se}F_{p3} - D$	0	0	0	0	0
$ut - T_4$	$K_{se}F$	0	0	0	0	0

Table 1.1: *(Continued.)* Coefficients of the terms in Eq. (1.41)

LHS $= \delta\Sigma$	$\exp\left(\frac{t_n-t}{\tau_{tac}}\right)$	$\exp\left(\frac{t_n-t}{\tau_{tde}}\right)$	$\exp\left(\frac{t_n-t}{\tau_{tn1}}\right)$	$\exp\left(\frac{t_n-t}{\tau_{tn2}}\right)$	$\exp\left(\frac{t_n-t}{\tau_{tn3}}\right)$
$u(t)$	0	$-\left(K_{se}-\frac{B_2}{\tau_{tde}}\right)K_{t0}$	$-\left(K_{se}-\frac{B_2}{\tau_{tn1}}\right)L_{t1}$	0	0
$ut-5\tau_{gn1}$	0	0	0	0	0
$ut-T_1$	0	0	0	0	0
$ut-T_2$	0	0	0	0	0
$ut-T_3$	$-\left(K_{se}-\frac{B_2}{\tau_{tac}}\right)K_{t1}$	$\left(K_{se}-\frac{B_2}{\tau_{tde}}\right)M_{t1}$	$\left(K_{se}-\frac{B_2}{\tau_{tn1}}\right)L_{t1}e^{-t_3/\tau_{tn1}}$	$-\left(K_{se}+\frac{B_2}{\tau_{tn2}}\right)L_{t2}e^{-5}$	0
$ut-T_3-5\tau_{tn2}$	$-\left(K_{se}-\frac{B_2}{\tau_{tac}}\right)M_{t2}$	0	0	$\left(K_{se}+\frac{B_2}{\tau_{tn2}}\right)L_{t2}$	0
$ut-T_4$	$-\left(K_{se}-\frac{B_2}{\tau_{tac}}\right)M_{t3}$	$-\left(K_{se}-\frac{B_2}{\tau_{tde}}\right)K_{t3}$	0	0	$-\left(K_{se}+\frac{B_2}{\tau_{tn3}}\right)L_{t3}$

$$D_4 = D_{gn1} = \frac{1}{\tau_{gn1}^3} + \frac{R_2}{\tau_{gn1}^2} + \frac{R_1}{\tau_{gn1}} + R_0;$$

$$D_5 = D_{gn2} = -\frac{1}{\tau_{gn2}^3} + \frac{R_2}{\tau_{gn2}^2} - \frac{R_1}{\tau_{gn2}} + R_0;$$

$$D_6 = D_{gn3} = -\frac{1}{\tau_{gn3}^3} + \frac{R_2}{\tau_{gn3}^2} - \frac{R_1}{\tau_{gn3}} + R_0;$$

$$D_7 = D_{tac} = -\frac{1}{\tau_{tac}^3} + \frac{R_2}{\tau_{tac}^2} - \frac{R_1}{\tau_{tac}} + R_0;$$

$$D_8 = D_{tde} = -\frac{1}{\tau_{tde}^3} + \frac{R_2}{\tau_{tde}^2} - \frac{R_1}{\tau_{tde}} + R_0;$$

$$D_9 = D_{tn1} = -\frac{1}{\tau_{tn1}^3} + \frac{R_2}{\tau_{tn1}^2} - \frac{R_1}{\tau_{tn1}} + R_0;$$

$$D_{10} = D_{tn2} = \frac{1}{\tau_{tn2}^3} + \frac{R_2}{\tau_{tn2}^2} + \frac{R_1}{\tau_{tn2}} + R_0;$$

$$D_{11} = D_{tn3} = -\frac{1}{\tau_{tn3}^3} + \frac{R_2}{\tau_{tn3}^2} - \frac{R_1}{\tau_{tn3}} + R_0.$$

Note that the signs for D_{gn1} and D_{tn2} are different from the others terms due to their positive expressions in the exponential term.

Natural Response   The first step in solving for the natural response, $\theta_n(t)$, is to determine the roots of the characteristic equation from Eq. (1.26), $s^3 + P_2 s^2 + P_1 s + P_0 = 0$. Using typical parameter values described later, the system has one real root, α_1, and one pair of complex roots, $\alpha_2 \pm j\beta_2$. Thus, using superposition with the delayed inputs, $\theta_n(t)$ is

$$\theta_n(t) = \left(K_{11}e^{\alpha_1 t} + K_{12}e^{\alpha_2 t}\cos(\beta_3 t) + K_{13}e^{\alpha_2 t}\sin(\beta_2 t) \right) u(t)$$

$$+ \left(\begin{array}{l} K_{21}e^{\alpha_1(t-5\tau_{gn1})} + K_{22}e^{\alpha_2(t-5\tau_{gn1})}\cos\left(\beta_3\left(t-5\tau_{gn1}\right)\right) \\ \\ + K_{23}e^{\alpha_2(t-5\tau_{gn1})}\sin\left(\beta_2\left(t-5\tau_{gn1}\right)\right) \end{array} \right) u\left(t-5\tau_{gn1}\right)$$

$$+ \left(\begin{array}{l} K_{31}e^{\alpha_1(t-5T_1)} + K_{32}e^{\alpha_2(t-5T_1)}\cos\left(\beta_3\left(t-5T_1\right)\right) \\ \\ + K_{33}e^{\alpha_2(t-5T_1)}\sin\left(\beta_2\left(t-5T_1\right)\right) \end{array} \right) u\left(t-5T_1\right)$$

$$+ \left(\begin{array}{l} K_{41}e^{\alpha_1(t-5T_2)} + K_{42}e^{\alpha_2(t-5T_2)} \cos\left(\beta_3\left(t-5T_2\right)\right) \\ +K_{43}e^{\alpha_2(t-5T_2)} \sin\left(\beta_2\left(t-5T_2\right)\right) \end{array} \right) (t-5T_2)$$

$$+ \left(\begin{array}{l} K_{51}e^{\alpha_1(t-5T_3)} + K_{52}e^{\alpha_2(t-5T_3)} \cos\left(\beta_3\left(t-5T_3\right)\right) \\ +K_{53}e^{\alpha_2(t-5T_3)} \sin\left(\beta_2\left(t-5T_3\right)\right) \end{array} \right) u(t-5T_3)$$

$$+ \left(\begin{array}{l} K_{61}e^{\alpha_1(t-T_3-5\tau_{tn2})} + K_{62}e^{\alpha_2(t-T_3-5\tau_{tn2})} \cos\left(\beta_3\left(t-T_3-5\tau_{tn2}\right)\right) \\ +K_{63}e^{\alpha_2(t-T_3-5\tau_{tn2})} \sin\left(\beta_2\left(t-T_3-5\tau_{tn2}\right)\right) \end{array} \right)$$
$$u(t-T_3-5\tau_{tn2})$$

$$+ \left(K_{71}e^{\alpha_1(t-T_4)} + K_{72}e^{\alpha_2(t-T_4)} \cos\left(\beta_3\left(t-T_4\right)\right) + K_{73}e^{\alpha_2(t-T_4)} \sin\left(\beta_2\left(t-T_4\right)\right) \right)$$
$$u(t-T_4),$$

$$(1.43)$$

where K_{ij} are the constants to be determined from the initial conditions.

Complete Response   The complete response, $\theta(t)$, is found by summing Eqs. (1.42) and (1.43), $\theta(t) = \theta_n(t) + \theta_f(t)$, giving

$$\theta(t) = \left(K_{11}e^{\alpha_1 t} + K_{12}e^{\alpha_2 t}\cos(\beta_3 t) + K_{13}e^{\alpha_2 t}\sin(\beta_2 t)u(t)\right)$$

$$+ \left(\begin{array}{l} K_{21}e^{\alpha_1(t-5\tau_{gn1})} + K_{22}e^{\alpha_2(t-5\tau_{gn1})}\cos\left(\beta_3\left(t-5\tau_{gn1}\right)\right) \\[2mm] + K_{23}e^{\alpha_2(t-5\tau_{gn1})}\sin\left(\beta_2\left(t-5\tau_{gn1}\right)\right) \end{array} \right) u(t-5\tau_{gn1})$$

$$\vdots$$

$$+ \left(\begin{array}{l} K_{71}e^{\alpha_1(t-t_4)} + K_{72}e^{\alpha_2(t-t_4)}\cos\left(\beta_3\left(t-t_4\right)\right) \\[2mm] + K_{73}e^{\alpha_2(t-t_4)}\sin\left(\beta_2\left(t-t_4\right)\right) \end{array} \right) u\left(t-t_4\right)$$

$$+ \delta\left(A_{1,1} + A_{1,2}e^{-t/\tau_{gac}} + A_{1,3}e^{-t/\tau_{gde}} + \ldots + A_{1,11}e^{-t/\tau_{tn3}}\right)u(t)$$

$$+ \delta\left(\begin{array}{l} A_{2,1} + A_{2,2}e^{-(t-5\tau_{gn1})/\tau_{gac}} + A_{2,3}e^{-(t-5\tau_{gn1})/\tau_{gde}} \\[2mm] + \ldots + A_{2,11}e^{-(t-5\tau_{gn1})/\tau_{tn3}} \end{array} \right) u\left(t-5\tau_{gn1}\right)$$

$$\vdots$$

$$+ \delta\left(\begin{array}{l} A_{7,1} + A_{7,2}e^{-(t-t_4)/\tau_{gac}} + A_{7,3}e^{-(t-t_4)/\tau_{gde}} \\[2mm] + \ldots + A_{7,11}e^{-(t-t_4)/\tau_{tn3}} \end{array} \right) u\left(t-t_4\right)$$

(1.44)

$$= \left(\begin{array}{l} K_{11}e^{\alpha_1 t} + K_{12}e^{\alpha_2 t}\cos(\beta_3 t) + K_{13}e^{\alpha_2 t}\sin(\beta_2 t) \\[2mm] + \delta\left(A_{1,1} + A_{1,2}e^{-t/\tau_{gac}} + A_{1,3}e^{-t/\tau_{gde}} + \ldots + A_{1,11}e^{-t/\tau_{tn3}}\right) \end{array} \right) u(t)$$

$$+ \left(\begin{array}{l} K_{21}e^{\alpha_1(t-5\tau_{gn1})} + K_{22}e^{\alpha_2(t-5\tau_{gn1})}\cos\left(\beta_3\left(t-5\tau_{gn1}\right)\right) \\[2mm] + K_{23}e^{\alpha_2(t-5\tau_{gn1})}\sin\left(\beta_2\left(t-5\tau_{gn1}\right)\right) \\[2mm] + \delta\left(\begin{array}{l} A_{2,1} + A_{2,2}e^{-(t-5\tau_{gn1})/\tau_{gac}} + A_{2,3}e^{-(t-5\tau_{gn1})/\tau_{gde}} \\[2mm] + \ldots + A_{2,11}e^{-(t-5\tau_{gn1})/\tau_{tn3}} \end{array} \right) \end{array} \right) u(t-5\tau_{gn1})$$

$$\vdots$$

$$+ \left(\begin{array}{l} K_{71}e^{\alpha_1(t-T_4)} + K_{72}e^{\alpha_2(t-T_4)}\cos\left(\beta_3\left(t-T_4\right)\right) \\[2mm] + K_{73}e^{\alpha_2(t-T_4)}\sin\left(\beta_2\left(t-T_4\right)\right) \\[2mm] \delta\left(\begin{array}{l} A_{7,1} + A_{7,2}e^{-(t-T_4)/\tau_{gac}} + A_{7,3}e^{-(t-T_4)/\tau_{gde}} \\[2mm] + \ldots + A_{7,11}e^{-(t-T_4)/\tau_{tn3}} \end{array} \right) \end{array} \right) u\left(t-T_4\right).$$

Next, we determine the unknown constants K_{11} to K_{73}, 21 terms in total. At the beginning of a saccade, the system is in steady-state and all initial conditions are assumed to be zero. Using the initial conditions, that is, $\theta(0) = 0$ (initial displacement), $\dot{\theta}(0) = 0$ (initial velocity) and $\ddot{\theta}(0) = 0$ (initial acceleration), and Eq. (1.44), we generate three equations to determine three of the unknown constants, K_{11}, K_{12}, and K_{13}, from

$$
\begin{aligned}
\theta(0) = 0 &= K_{11} + K_{12} + A_{1,1} + A_{1,2} + \ldots + A_{1,11}, \\
\dot{\theta}(0) = 0 &= \alpha_1 K_{11} + \alpha_2 K_{12} + \beta_2 K_{13} - \left(\frac{A_{1,2}}{\tau_{gac}} + \ldots + \frac{A_{1,11}}{\tau_{tn3}} \right), \\
\ddot{\theta}(0) = 0 &= \alpha_1 K_{11} + \left(\alpha_2^2 - \beta a_2^2 \right) K_{12} + 2\alpha_2 \beta_2 K_{13} + \frac{A_{1,2}}{\tau_{gac}^2} + \ldots + \frac{A_{1,11}}{\tau_{tn3}^2}.
\end{aligned}
\tag{1.45}
$$

The other unknown constants, K_{21} to K_{73}, are determined at each of the switch times, based on the fact that the variables $\theta(t)$, $\dot{\theta}(t)$, and $\ddot{\theta}(t)$ must be continuous in time. For instance, at switch time T_1, $\theta(T_1^-) = \theta(T_1^+)$, $\dot{\theta}(T_1^-) = \dot{\theta}(T_1^+)$ and $\ddot{\theta}(T_1^-) = \ddot{\theta}(T_1^+)$. Thus, at time T_1 we have

$$
\begin{aligned}
\theta(T_1) &= K_{11} e^{\alpha_1 T_1} + \ldots + A_{2,11} e^{-(T_1 - 5\tau_{gn1})/\tau_{tn3}} \\
&= K_{11} e^{\alpha_1 T_1} + \ldots + A_{2,11} e^{-(T_1 - 5\tau_{gn1})/\tau_{tn3}} \\
&\quad + K_{31} + K_{32} + A_{3,1} + A_{3,2} + \ldots + A_{3,11} \\[2mm]
\dot{\theta}(T_1) &= \alpha_1 K_{11} e^{\alpha_1 T_1} + \ldots - \frac{A_{2,11} e^{-(T_1 - 5\tau_{gn1})/\tau_{tn3}}}{\tau_{tn3}} \\
&= \alpha_1 K_{11} e^{a_1 T_1} + \ldots - \frac{A_{2,11} e^{-(T_1 - 5\tau_{gn1})/\tau_{tn3}}}{\tau_{tn3}} + \alpha_1 K_{31} + \alpha_2 K_{32} \\
&\quad + \beta_2 K_{33} - \left(\frac{A_{3,2}}{\tau_{gac}} + \ldots + \frac{A_{3,11}}{\tau_{tn3}} \right) \\[2mm]
\ddot{\theta}(T_1) &= \alpha_1^2 K_{11} e^{\alpha_1 T_1} + \ldots + \frac{A_{2,11} e^{-(T_1 - 5\tau_{gn1})/\tau_{tn3}}}{\tau_{tn3}^2} \\
&= \alpha_1^2 K_{11} e^{a_1 T_1} + \ldots + \frac{A_{2,11} e^{-(T_1 - 5\tau_{gn1})/\tau_{tn3}}}{\tau_{tn3}^2} + \alpha_1 K_{31} + \alpha_2^2 - \alpha_2^2 K_{32} \\
&\quad + 2\alpha_2 \beta_2 K_{33} + \frac{A_{32}}{\tau_{gac}^2} + \ldots + \frac{A_{3,11}}{\tau_{tn3}^2},
\end{aligned}
\tag{1.46}
$$

which requires

$$0 = K_{31} + K_{32} + A_{3,1} + A_{3,2} + \ldots + A_{3,11}$$
$$0 = \alpha_1 K_{31} + \alpha_2 K_{32} + \beta_2 K_{33} - \left(\frac{A_{3,2}}{\tau_{gac}} + \ldots + \frac{A_{3,11}}{\tau_{tn3}} \right) \qquad (1.47)$$
$$0 = \alpha_1 K_{31} + \alpha_2^2 - \beta_2^2 K_{32} + 2\alpha_2 \beta_2 K_{33} + \frac{A_{32}}{\tau_{gac}^2} + \ldots + \frac{A_{3,11}}{\tau_{tn3}^2}.$$

The constants K_{31}, K_{32}, and K_{33} are solved from Eq. (1.47). The equations for the other unknown constants at the switch times follow similarly.

Example 1.1   Problem

Consider the oculomotor system shown in Fig. 1.1. Given the initial conditions and parameter below, create a Simulink program and plot the neural inputs, actives state tensions, position, velocity, and acceleration.

$\theta(0) = 0°, \dot{\theta}(0) = 0°\text{s}^{-1}, \ddot{\theta}(0) = \ddot{0}°\text{s}^{-2}, T_1 = 0.0044 \text{ s}, T_2 = 0.0259 \text{ s}, T_3 = 0.0293 \text{ s}, T_4 = 0.0462 \text{ s}, F_{p1} = 1.06\text{N}, F_{p2} = 0.9331\text{N}, F_{p3} = 0.3790\text{N}, F_{g0} = 0.4\text{N}, F_{gs} = 0.5546\text{N}, F_{t0} = 0.4\text{N}, F_{ts} = 0.2895\text{N}, \tau_{gn1} = 0.000287\text{s}, \tau_{gn2} = 0.0034\text{s}, \tau_{gn3} = 0.0042\text{s}, \tau_{gac} = 0.0112\text{s}, \tau_{tn1} = 0.000939\text{s}, \tau_{tn2} = 0.0012\text{s}, \tau_{tn3} = 0.001\text{s}, \tau_{tac} = 0.0093\text{s}, \tau_{tde} = 0.0048\text{s}, K_{se} = 124.9582\text{Nm}, K_{lt} = 60.6874\text{Nm}, K = 16.3597\text{Nm}, B_1 = 5.7223\text{Nms}^{-1}, B_2 = 0.5016\text{Nms}^{-1}, B = 0.327\text{Nms}^{-1}, J = 0.0022\text{Nms}^{-1},$ and radius $= 0.0118$ m.

Solution

We first compute the following intermediate results

```
b12=b1+b2=6.2239
kst=kse+klt=185.6456
c3=b12*j=0.0137
c2=b12*bp+kst*j+2*b1*b2=8.1855
c1=b12*kp+kst*bp+2*(b2*klt+b1*kse)= 1.6535e+003
c0=kst*kp+2*kse*klt=1.8204e+004
delta=57.296/(r*c3)= 3.5397e+005
p2=c2/c3=596.7159
p1=c1/c3=1.2054e+005
p0=c0/c3=  1.3271e+006
```

The Simulink model is shown in Fig. 1.6. The diagram in (A) is implemented using Eq. (1.26). The input to the oculomotor plant is shown in diagram (B), with the agonist and antagonist active-state tensions shown in (C) and (D) based on Eqs. (1.29) and (1.30). The neural input is based on Eqs. (1.27) and (1.28).

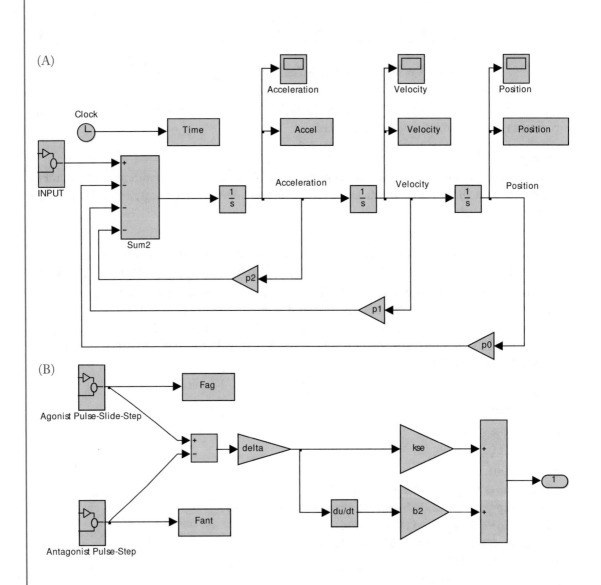

Figure 1.6: Simulink program for Example 1.1. (A) Main program. (B) Input to plant.

(C)

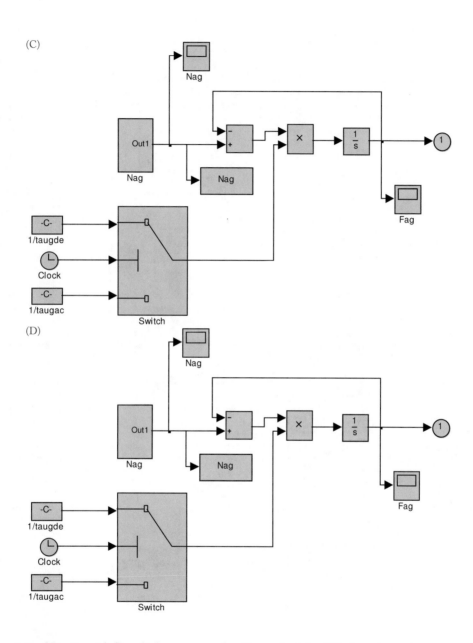

(D)

Figure 1.6: *(Continued.)* Simulink program for Example 1.1. (C) Agonist active-state tension. (D) Antagonist active-state tension.

(E)

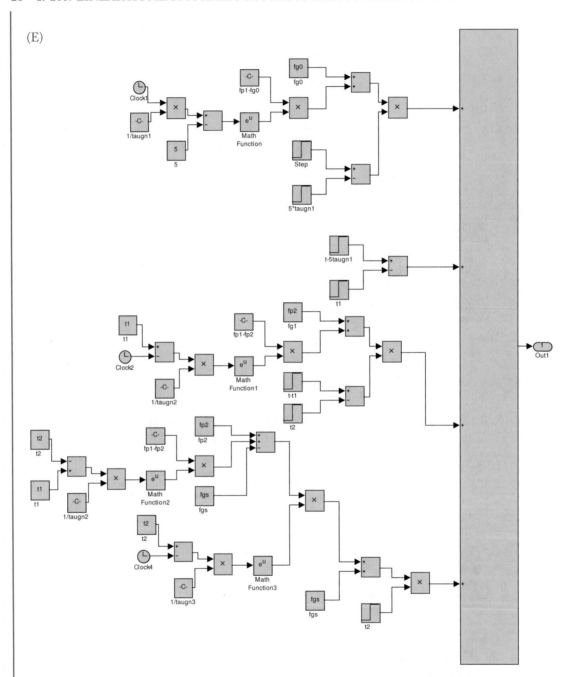

Figure 1.6: *(Continued.)* Simulink program for Example 1.1. (E) Agonist neural input.

(F)

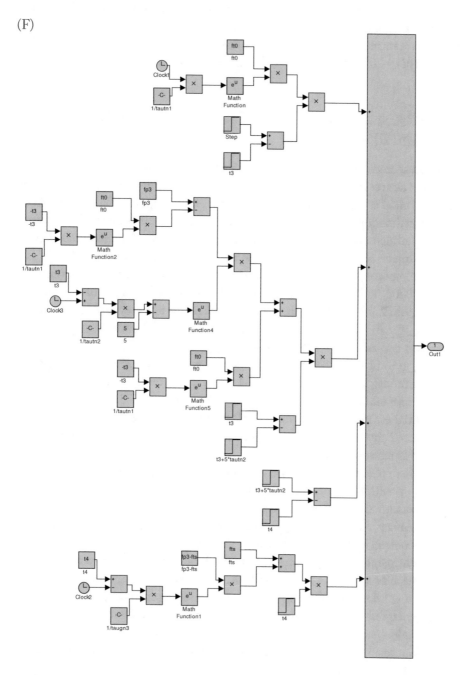

Figure 1.6: *(Continued.)* Simulink program for Example 1.1. (F) Agonist neural input.

In Fig. 1.7 are plots of position, velocity, acceleration, agonist neural input and active-state tension, and antagonist neural input and active-state tension. From the antagonist neural input and active-state tension plot, it is clear that the eye movement has post-saccade behavior. Peak return velocity is $-20° \text{ s}^{-1}$ in Fig. 1.7 (B), which makes it a glissade.

1.3   PARAMETER ESTIMATION AND SYSTEM IDENTIFICATION

The model presented here involves a total of 25 parameters describing the oculomotor plant, neural inputs, and active-state tensions that are estimated by system identification as described in the next section. Initial estimates of the model parameters are important since they affect the convergence of the estimation routine. In this model, the initial estimates are derived from previously published experimental observations and briefly reviewed here. A more detailed discussion of the parameter estimates for human and monkey are given in Sections 1.4 and 1.5.

The oculomotor plant parameters for a human are determined based on the work by Enderle et al. (1988a, 2002, 1991) and Zhou et al. (2009). The eyeball moment of inertia, J, has an initial estimated value of $2.2 \times 10^{-3} \text{ Ns}^2/\text{m}$, assuming the radius of the eyeball is 11 mm. The parameter values of K_{se}, K_{lt}, B_1, B_2 are based on a linear muscle model (Enderle et al., 1991). From the length-tension curves, $K_{se} = 125 \text{ N/m}$, and $K_{lt} = 60.7 \text{N/m}$. B_1 is selected as 5.6 Ns/m, and B_2 is 0.5 Ns/m to fit the nonlinear force-velocity relationship. K is determined by steady-state analysis of the model, yielding a value of 16.34 N/m. B is determined by considering the dominant orbital time constant 0.02 s, which yields $B = 0.327 \text{ Ns/m}$.

The initial estimated start time, T_p, of the saccade is based on selecting a threshold from the estimated velocity from the position data. Such estimations can also be obtained using a first-spike method as reported by others.

The nine parameters for the agonist neural input and active-state tension are T_1, T_2, F_{p1}, F_{p2}, with time constants $\tau_{gn1}, \tau_{gn2}, \tau_{gn3}, \tau_{gac}$, and τ_{gde}. The initial estimate of the duration, T_1, is assumed to be 3 ms. F_{p2}, τ_{gn2} and τ_{gn3} are empirically estimated from EBN firing rates of the monkey. F_{p1}, τ_{gac}, and $T_2 - T_1$ are estimated from the peak velocity of the data, using the method reported by Enderle and Wolfe (1988a).

The eight parameters for antagonist neural input and active-state tension are T_3, T_4, F_{p3}, with time constants $\tau_{tn1}, \tau_{tn2}, \tau_{tn3}, \tau_{tac}$, and τ_{tde}. The initial estimates for time constants τ_{tn1}, τ_{tn2}, and τ_{tn3} are 2 ms, 1.5 ms, and 0.2 ms, respectively. The antagonist onset delay, $T_3 - T_2$, is the time between the antagonist step and agonist pulse, and it is variable from saccade to saccade. Physiological observations suggest that the delay varies from 3 ms to 20 ms for large saccades (Robinson, 1981). Without losing generality, the antagonist onset time, T_3, is estimated as a time between T_2 and a time near the end of the saccade.

A second peak velocity is observed in saccades with either a dynamic or glissadic overshoot. Data suggest that dynamic overshoot has a higher second peak velocity than glissadic overshoot.

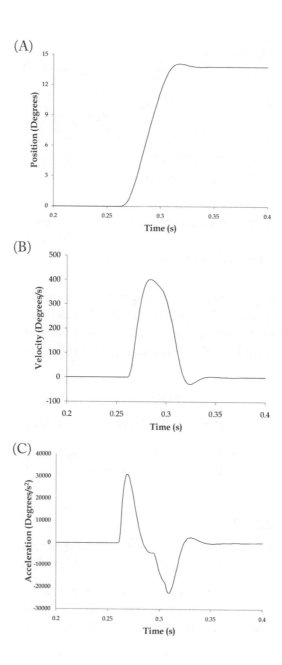

Figure 1.7: Plots of position (A), velocity (B), and acceleration (C) for Example 1.1.

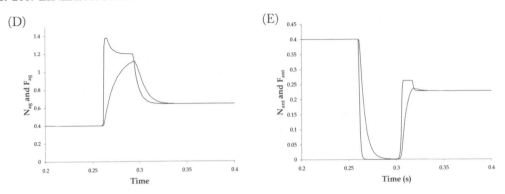

Figure 1.7: *(Continued.)* Plots of agonist neural input and active-state tension (D) and antagonist neural input and active-state tension (E) for Example 1.1. Active-state tension is drawn with a blue line and neural input is drawn with a red line.

In fact, we show that the prominent second peak velocity may be caused by the antagonist PIRB. The antagonist parameters F_{p3}, T_4, and τ_{tac} control the rebound burst and are determined by analyzing the relationships between them and the second peak velocity.

Steady-state tensions F_{g0}, F_{gs}, F_{t0}, and F_{ts}, are determined as functions of eye position at steady-state

$$
F = \begin{cases} 0.4 + 0.0175\,|\theta|\ N & \text{for } \theta \geq 0° \\ 0.4 - 0.0125\,|\theta|\ N & \text{for } \theta < 0° \end{cases}
$$

where the eye position θ is always positive for the agonist and negative for the antagonist active-state tension (Enderle et al., 2005). These four parameters are fixed for a particular saccade during the estimation routine.

1.3.1 SYSTEM IDENTIFICATION

In a time domain system identification method, a conjugate gradient descent algorithm is applied directly to the objective function. The system identification problem is stated as follows:

Find a parametric vector $\mathbf{y} = (y_1, y_2, \ldots, y_n)$ of the oculomotor plant to minimize (optimize) an objective function $f = f(\mathbf{y})$ subject to equality constraints $h_i(\mathbf{y}) = 0, i = 1\ldots q$ and inequality constraints $g_i(\mathbf{y}) \leq \mathbf{0}, i = 1, \ldots, m$.

Here, the parametric vector $\mathbf{y}$ is composed of the 25 parameters that need to be estimated. Let $\theta_M(\mathbf{y}, t)$ be the solution of the homeomorphic oculomotor plant as defined by Eq. (1.26). The analytical form of $\theta_M(\mathbf{y}, t)$ is derived by substituting the analytical solutions of active-state tensions and using the separation of homogeneous and particular solutions.

Next, let $p_1, \ldots p_N$ be a set of experimental data points of eye position in time domain $p_i = (t_i, \theta_i)$, where t_i is the ith time point and θ_i is the ith eye position in degrees. The 2009 linear homeomorphic model is defined by its own parametric vector $\mathbf{a} = (P_2, P_1, P_0, \delta, K_{se}, B_2, F_{ag}, F_{ant})$

$$f(P, \mathbf{a}) = \dddot{\theta} + P_2 \ddot{\theta} + P_1 \dot{\theta} + P_0 \theta - \delta \left(K_{se} \left(F_{ag} - F_{ant} \right) + B_2 \left(\dot{F}_{ag} - \dot{F}_{ant} \right) \right) = 0 \qquad (1.48)$$

where $P = \bigcup_{i \in H} p_i$, which includes all possible time-position points during saccade.

The goal is to find $\mathbf{x}^*$ that satisfies

$$\mathbf{x}^* = \min \arg \sum_{i=1}^{N} [D(p_i, \mathbf{a})]^2, \qquad (1.49)$$

where $D(p_i, \mathbf{a})$ is a suitable distance.[2] Here the distance is defined as the algebraic distance of experimental data point p_i from the model $|f(p_i, \mathbf{a})|$. Thus, Eq. (1.49) becomes

$$\mathbf{x}^* = \min \arg \sum_{i=1}^{N} |f(p_i, \mathbf{a})|^2. \qquad (1.50)$$

The vector $\mathbf{a}$ in Eq. (1.50) is then represented by the parameter vector $\mathbf{x}$, and thus the objective function is

$$F(\mathbf{x}) = \sum_{i=1}^{N} |f(p_i, \mathbf{x})|^2, \qquad (1.51)$$

subject to:

1. *Equality constraints*

   The initial and final positions should satisfy

   $$h_1 = K_{se}(F_{ag}(t = 0) - F_{ant}(t = 0)) - P_0 \theta(t = 0) = 0,$$
   $$h_2 = K_{se}(F_{ag}(t = t_s) - F_{ant}(t = t_s)) - P_0 \theta(t = t_s) = 0,$$

   where at time t_s the saccade ends.

2. *Inequality constraints*

   For parameters $(K_{se}, K_{lt}, B_1, B_2, K_p, B_p, J)$, inequality constraints are added to limit variations within a desirable range around their initial estimations. Those estimations are obtained from experimental data. The purpose of the inequality constraints is to eliminate the

---

[2]In the frequency method, this distance is defined as error squared between transfer function $G(jw, b)$ and the plant frequency response $A(w)e^{j\theta(w)}$ as carried out in Enderle (1988b).

possibility of abnormal values from the gradient method. Here, we set a constraint within 25% around the initial guess

$$
\begin{cases}
g_i & = x_i/x_i^0 - (1 + 25\%) \leq 0 \\
g_{i+1} & = (1 - 25\%) - x_i/x_i^0 \leq 0,
\end{cases}
$$

for each parameter listed above. This constrained optimization problem is transformed into an unconstrained problem using a transformation function of the form:

$$
\phi(\mathbf{x}, \mathbf{r}) = F(\mathbf{x}) + P(\mathbf{h}(\mathbf{x}), \mathbf{g}(\mathbf{x}), \mathbf{r}), \tag{1.52}
$$

where $\mathbf{r}$ is a vector of penalty parameters and P is a real valued function whose action is imposing the penalty on the objective function controlled by $\mathbf{r}$. P is defined by the following quadratic loss function:

$$
P(\mathbf{h}(\mathbf{x}), \mathbf{g}(\mathbf{x}), \mathbf{r}) = r_1 \sum_{i=1}^{p} [h_i(\mathbf{x})]^2 + r_2 \sum_{i=1}^{m} \log(-g_i(\mathbf{x})). \tag{1.53}
$$

Here, the log barrier function methods are applied to the inequality constraints. In fact, the function becomes infinite if any of the inequalities are active. When the iterative process is started from a feasible point (initial guess), it cannot go into the infeasible region because the iterative process cannot cross the barrier. If $h_i(\mathbf{x}) \neq 0$, Eq. (1.53) also gives a positive value to the function P, and the cost function is penalized.

1.3.2   NUMERICAL GRADIENT

One shortcoming of the time domain method is that the exact gradient cannot be evaluated as in the frequency response method (Enderle and Wolfe, 1988a). This is because the parameter vector in the objective function, Eq. (1.49), is implicit and the derivatives are impossible to calculate directly.

At first glance, a simple one-step 1st- or 2nd-order finite differencing method appears plausible. However, due to truncation and round-off error in Taylor series expansion and the parameters with different scales, this method makes the parameter values unstable and without convergence. Here, the finite differencing is calculated with smaller and smaller finite values of the step h, to $h \rightarrow 0$. By the use of Neville's algorithm, each finite differencing calculation produces both an extrapolation of higher order and extrapolations of previous lower orders, but with smaller scales h.

The algorithm is based on Ridders, with some necessary changes. The derivatives are evaluated in the objective function with respect to each parameter individually. The initial step length is estimated by

$$
h = eps * \max(|parm|, 1/s) * sign(parm) \tag{1.54}
$$

where s is a large number (i.e., 1000) and eps is a small number.

1.3.3   VELOCITY AND ACCELERATION ESTIMATION

Velocity and acceleration are essential in the study of the oculomotor system. They are computed from data by the central difference method. To reduce aliasing, a suitable spread for central differencing must be estimated. Based on the relationship between the bandwidth and the amount of spread between points,

$$\text{bandwidth (Hz)} = \frac{0.443 \times \text{sampling rate (Hz)}}{\text{spread}}.$$

The maximum frequency for saccade velocity is estimated at 74 Hz, and 45 Hz for saccade acceleration (Bahill et al., 1980). Thus, for the saccade that has a sampling rate of 1000 Hz, the spread for saccade velocity is 6, and the spread for saccade acceleration is estimated as 8. The velocity is calculated as

$$\dot{y}(kT) = \frac{y((k+3)T) - y((k-3)T)}{6T},$$

which has a bandwidth of 74 Hz. The acceleration is estimated as

$$\ddot{y}(kT) = \frac{\dot{y}((k+4)T) - \dot{y}((k-4)T)}{8T},$$

which has a bandwidth of 55 Hz.

For the data sampled at 2000 Hz, the spread for saccade velocity is 12 and the spread for saccade acceleration is estimated as 16. The velocity is estimated as

$$\dot{y}(kT) = \frac{y((k+6)T) - y((k-6)T)}{12T},$$

which has a bandwidth of 74 Hz. And the acceleration is estimated as

$$\ddot{y}(kT) = \frac{\dot{y}((k+8)T) - \dot{y}((k-8)T)}{16T},$$

which has a bandwidth of 55 Hz.

1.3.4   INVERSE FILTER

The inverse filter is a discrete first-order low-pass filter (LPF) that filters the neuron bursting rate to active-state tension. Consider the EBN firing rate, where in the continuous time domain, the LPF has the form of

$$\dot{F}_{ag} = \frac{N_{ag} - F_{ag}}{\tau_{ag}}, \text{ or } \tau_{ag}\dot{F}_{ag} + F_{ag} = N_{ag}. \tag{1.55}$$

Since the EBN firing rate is discontinuous, the differential equation can be discretized using the approximation:

$$\frac{dF_{ag}(t)}{dx} \approx \frac{F_{ag,k} - F_{ag,k-1}}{T_s}, \tag{1.56}$$

where T_s is the interval between each measurement, i.e., the sampling interval 0.0005 s at frequency of 2000 Hz. Thus, the differential equation representing the 1st-order low pass filter is converted to

$$\tau_{ag} \frac{F_{ag,k} - F_{ag,k-1}}{T_s} + F_{ag,k} = N_{ag,k}. \tag{1.57}$$

Simplification and re-arrangement of Eq. (1.57) gives

$$F_{ag,k} = \left(\frac{\tau_{ag}}{\tau_{ag} + T_s} \right) F_{ag,k-1} + \left(\frac{T_s}{\tau_{ag} + T_s} \right) N_{ag,k}. \tag{1.58}$$

Here, $F_{ag,k}$ and $F_{ag,k-1}$ are the agonist active-state tensions at time point k and $k-1$, respectively. $N_{ag,k}$ is the burst frequency at sampling time point k, T_s is the sampling interval, and $\tau_{ag} = \tau_{gac}(u(t) - u(t - t_2)) + \tau_{gde}u(t - t_2)$.

We transform the neuron firing rate in Hz to active-state tension in N with the following normalization

$$\frac{H - H_0}{F - F_{g0}} = S \quad \left(\frac{Hz}{N} \right), \tag{1.59}$$

where H is the firing rate, H_0 is the steady-state firing rate before the saccade, F is the active-state tension, F_{g0} is the initial active-state tension, and S is the coefficient in $\left(\frac{Hz}{N} \right)$. This equation scales the low-pass filtered neuron firing rate to an active-state tension. A coefficient selected based on physiological evidence is

$$S = \frac{100}{0.4}, \tag{1.60}$$

which is a reasonable approximation (Enderle, 2002). Practically, S and H_0 are quite variable for saccades of different sizes, and are therefore selected manually to match the data. This is because the firing data from a single neuron may not represent the average firing of all the neurons during a saccade. As a result, the observation from the firing of a single or several neurons is not accurate for general use in the generation of saccades.

1.4   INITIAL PARAMETER ESTIMATION FOR HUMANS

Suitable initial estimates for model parameters are important for system identification accuracy. To meet local convergence requirements of conjugate gradient algorithm, a reasonable initial guess is required for a success in the iteration process. This section describes the algorithms used in calculating initial parameter estimates.

1.4.1   ESTIMATION OF THE START TIME AND DURATION OF A SACCADE

As a first step, the start time and duration of a saccade are estimated by using thresholds. The velocity profile of a saccade with glissadic overshoot is shown in Fig. 1.8. To avoid the detection error due to noise before filtering, *Threshold 1* is set at a relatively large value (i.e., 70°s^{-1}). The estimate of the saccade start time is 6 ms before the time when the velocity reaches *Threshold 1*. *Threshold 2* represents the end of the saccade and is set at $7°s^{-1}$. The saccade end time is 6 ms after the time when the velocity falls below *Threshold 2*.

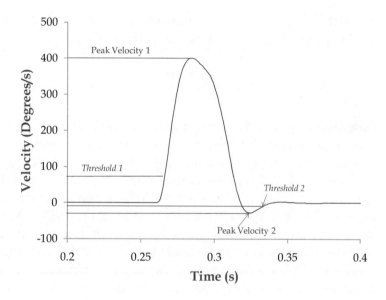

Figure 1.8: Velocity profile for a typical saccade with glissadic overshoot.

Obviously, these two thresholds are artificially specified and do not have precise relationships with the real start time and end time of the saccade. These estimates are required since the Kaiser window filter used in the estimation routine needs them. After filtering, these two parameters are evaluated again. The saccade start time is chosen as a time when the velocity is nearing $0°s^{-1}$ and is about to increase. The saccade end time is when the agonist and antagonist active-state tensions reach steady-state at their final values.

1.4.2   ESTIMATION OF MODEL PARAMETERS

The fixed parameters for the model are determined first and held constant for all saccades across all subjects. J has a value of $2.2 \times 10^{-3} \frac{Ns^2}{m}$ when the radius of the eyeball is 11 mm. From the length-tension curves, $K_{se} = 125\frac{N}{m}$ and $K_{lt} = 60.7\frac{N}{m}$ (Enderle et al., 1991). K is determined by steady-state analysis of the model as described using Eqs. (1.61) and (1.62). At the beginning of saccade, Eq. (1.26) is

$$K_{se}(F_{g0} - F_{t0}) = P_0\theta\,(0)$$
$$= (K_{st}K + 2K_{lt}K_{se})\theta\,(0)\,, \tag{1.61}$$

and at the end of saccade

$$K_{se}(F_{gs} - F_{ts}) = P_0\theta(+\infty)$$
$$= (K_{st}K + 2K_{lt}K_{se})(+\infty).$$

(1.62)

K should satisfy both Eqs. (1.61) and (1.62), yielding a value of $16.34\frac{N}{m}$. The initial estimate for B is determined by using the orbital time constant of 0.02 s, which is

$$\frac{B}{K} = 0.02s.$$

(1.63)

Thus, $B = 0.327\frac{Ns}{m}$.

B_1 is a dominant parameter that strongly influences the peak velocity of saccades. For the input, the dominant parameters that affect the peak velocity are F_{p1} and τ_{gac}. The influence of F_{p1} and τ_{gac} on the peak velocity is plotted in Fig. 1.9 (A) τ_{gac} and (B) F_{p1}. Figure 1.9 (A) illustrates the influence of B_1 and τ_{gac} on peak velocity with F_{p1} held constant. B_1 has a negative influence on peak velocity, while τ_{gac} has a positive influence on peak velocity. Figure 1.9 (B) shows the influence of B_1 and F_{p1} on peak velocity when τ_{ac} is 0.012 s. As before, B_1 has a negative influence on peak velocity, while F_{p1} has a positive influence on peak velocity. Since the typical peak velocities for saccades range from $200°s^{-1}$ to $800°s^{-1}$, B_1 is selected as 5.6 Ns/m and B_2 is selected as 0.5 Ns/m. Several force-velocity curves for different values of B_1 are shown in Fig. 1.10. They maintain a non-linear shape for the range of values from 2 to 6 since B_2 determines this property.

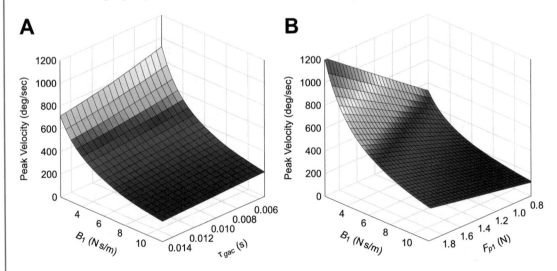

Figure 1.9: Influence of B_1, F_{p1} and τ_{gac} on the peak velocity of saccade. (A) F_{p1} is held constant at 1.2 N; (B) τ_{gac} is held constant at 0.012 s. (Note that F_{p2} is set equal to F_{p1}.)

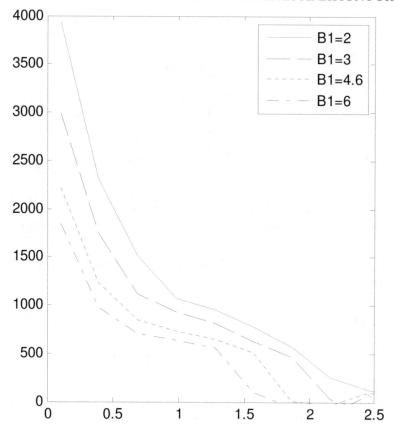

Figure 1.10: Force-velocity relationships for different values of B_1.

1.4.3   ESTIMATION OF PARAMETERS FOR THE AGONIST MUSCLE

The parameters for agonist muscles are T_1, T_2, F_{p1}, and F_{p2}, and time constants $\tau_{gn1}, \tau_{gn2}, \tau_{gn3}, \tau_{gac}$, and τ_{gde}. The EBN burst firing has a minimum duration T_1 for saccades for all sizes (Enderle, 2002). In the model, this duration is assumed to be 3 ms. F_{p2}, τ_{gn2} and τ_{gn3} are empirically estimated from observations of the EBN firing rate in monkeys. F_{p2} is estimated as a function of F_{p1},

$$F_{p2} = F_{gs} + 0.75 \left(F_{p1} - F_{gs} \right). \tag{1.64}$$

τ_{gn2} and τ_{gn3} are estimated from the shape of the firing rate and assumed to be dependent on the saccade magnitude. This dependency may be related to the different durations for saccades of different sizes.

From Fig. 1.11, the empirical equation for τ_{gn2} is

$$\tau_{gn2} = \min(0.00039 + 0.00034 \, |\theta| \, , 0.007) \tag{1.65}$$

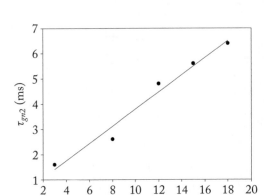
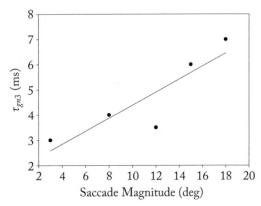

Figure 1.11: τ_{gn2} and τ_{gn3} are assumed to be dependent on the saccade magnitude. The lines are a linear regression result from the data points.

and the equation for τ_{gn3} is

$$\tau_{gn3} = \min(0.0018 + 0.00026 |\theta|, 0.007). \tag{1.66}$$

F_{p1}, τ_{gac}, and $T_2 - T_1$ are estimated from peak velocity (the primary Peak Velocity 1 in Fig. 1.8). The time constants for antagonist τ_{tn1} and τ_{tde} also influence the peak velocity. We have assumed them to be 0.002 s and 0.005 s, respectively. Since T_3 is assumed to always be larger than T_2, the other antagonist parameters do not affect the peak velocity.

The time to peak velocity is estimated from

$$\left. \frac{\partial^2 \theta}{\partial t^2} \right|_{t=T_{mv}} = 0, \tag{1.67}$$

where at $t = T_{mv}$, the peak velocity is obtained. As an example, Fig. 1.12 illustrates the affect of agonist pulse magnitude, F_{p1}, and duration, $T_2 - T_1$, on Peak Velocity 1 under different activation time constants, τ_{ac}, for an adducting 8° saccade. In this figure, the range of F_{p1} comes from the fact that the typical range of the peak firing rate of motoneuron is 200–600 Hz. Accordingly, with 0.4 N $\sim$ 100 Hz, the range of F_p is typically from 0.8 N to 3 N.

T_{mv} for the saccade is around 16 ms. Increasing the agonist pulse magnitude increases Peak Velocity 1. Increasing pulse duration increases Peak Velocity 1 until $T_{mv} - T_1$, after which the duration does not have much influence on peak velocity. Note that, due to the existence of minimum pulse duration T_1, the saccade could always achieve a velocity larger than $100°\text{s}^{-1}$. Increasing the agonist activation time constant, τ_{gac}, decreases the peak velocity as illustrated in Fig. 1.13. As expected, increasing the agonist activation time constant, τ_{gac}, decreases the peak velocity.

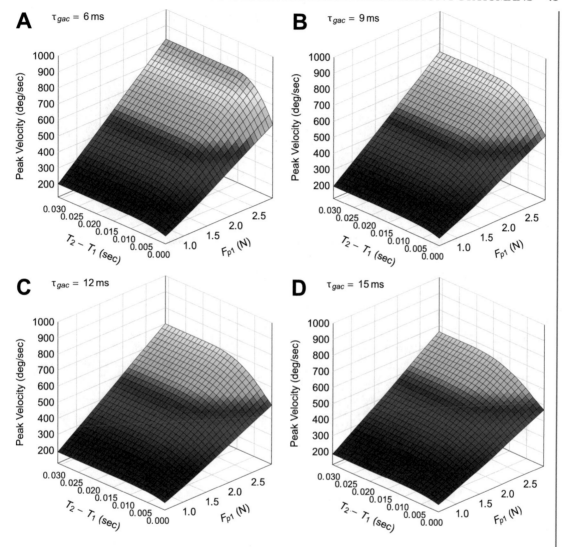

Figure 1.12: The effect of the agonist pulse magnitude, F_{p1}, and duration, $T_2 - T_1$, on Peak Velocity 1 using a different activation time constant, τ_{gac}.

These observations provide three clues for estimating agonist pulse magnitude, F_{p1}, duration, $T_2 - T_1$, and the activation time constant, τ_{gac}:

- Pulse magnitude strongly influences peak velocity. Peak velocity increases while F_{p1} increases.

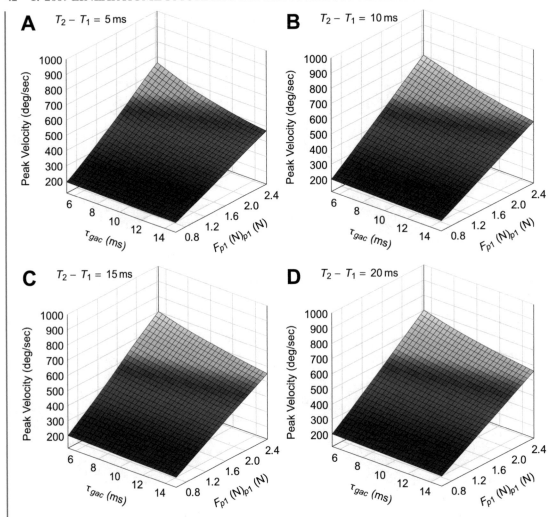

Figure 1.13: The effect of the agonist pulse magnitude F_{p1} and activation time constant τ_{gac} on Peak Velocity 1 using different durations $T_2 - T_1$.

- When duration $T_2 - T_1$ is larger than the threshold, it shows insignificant influences on peak velocity. The duration of the agonist pulse is important for reaching the expected displacement of saccade.

- Increasing the time constant τ_{gac} decreases the peak velocity.

According to the dynamic characteristics of the system and these observations, it is reasonable to roughly estimate T_2 equal to T_{mv} before final corrections. Due to the linear characteristic of the model, T_2 could be smaller or larger than T_{mv}. For a large saccade (i.e., the saccade magnitude is larger than 10°), T_2 is often larger than T_{mv} as observed in the data. T_2 can also be larger than T_{mv}, that being necessary to reach the saccade destination without changing peak velocity.

Based on these observations, the agonist pulse magnitude F_{p1} and time constant τ_{gac} are determined by setting up a $F_{p1} - \tau_{gac}$ versus peak velocity look-up table, which comes from the values of the nodes on the map shown in Fig. 1.13, and finding the most suitable Peak Velocity 1 in this table to fit the experimental data in the acceptable range of estimated peak acceleration. This matches the peak velocity of the model to that of the data. The time constant τ_{gn1} also has some influence on peak velocity; however, it primarily affects the peak acceleration.

As shown in Fig. 1.14, the agonist time constant, τ_{gn1}, is estimated from a look-up table to best match the peak acceleration. For the agonist, the decaying time constant, τ_{gde}, is assumed as 0.0065 s. Further, this parameter does not influence the time to the peak velocity or the peak velocity.

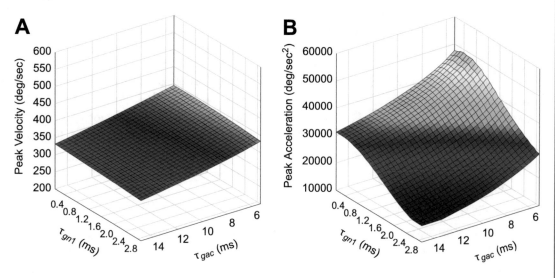

Figure 1.14: The effect of the agonist time constant τ_{gn1} and activation time constant τ_{gac} on (A) Peak Velocity 1 and (B) peak acceleration.

1.4.4   ESTIMATION OF PARAMETERS FOR ANTAGONIST MUSCLE

The parameters for the antagonist muscle are T_3, T_4, and F_{p3}, and time constants, τ_{tn1}, τ_{tn2}, τ_{tn3}, τ_{tac}, and τ_{tde}. The time constants τ_{tn1}, τ_{tn2}, and τ_{tn3} are estimated with values of 2 ms, 1.5 ms, and 0.2 ms, respectively. The antagonist onset delay $T_3 - T_2$ is the delay between antagonist step (or rebound burst) and agonist pulse. This delay is variable for saccades. Physiological observations suggest that the delay varies from 3 ms to as long as 20 ms for large saccades (Robinson, 1981). Without losing generality, the antagonist onset time T_3 is estimated as a time between T_2 and the time point near the end of saccade. This value is adjusted in the correction subroutine of the initial estimation routine.

A second peak velocity (Peak Velocity 2 in Fig. 1.8) is observed in many saccades, those with dynamic and glissadic overshoot. Data suggest that saccades with dynamic overshoot often have a higher Peak Velocity 2 than saccades with glissadic overshoot. In fact, the prominent second peak velocity mainly results from the antagonist rebound burst pulse. The agonist pulse duration has little influence on this peak velocity. The antagonist parameters F_{p3}, T_4, and τ_{tac} control the rebound burst and are determined by analyzing the relationships among them and Peak Velocity 2. Assuming that τ_{tde} has little influence on the magnitude of Peak Velocity 2, Peak Velocity 2 is plotted in Fig. 1.15 under the influence of F_{p3}, T_4, and τ_{tac}.

Fig. 1.15 shows that τ_{tac} has a negative influence on the Peak Velocity 2. The rebound burst pulse magnitude F_{p3} strongly affects Peak Velocity 2. Peak Velocity 2 increases when the duration increases until saturation. As a matter of fact, since the second peak velocities are mostly under $-40°s^{-1}$ for saccades, the influence of τ_{tac} is limited to the range $(-40$ to $0°s^{-1})$ and saturation is hardly achieved. In this range, a look-up table for F_{p3} and the rebound burst duration is used for Peak Velocity 2; one pair of F_{p3} and duration is selected from this table to match the Peak Velocity 2.

However, it should be noted that the Peak Velocity 2 is often small and is easily contaminated by noise. In some cases, a normal saccade can have a "fake" Peak Velocity 2, which is, in fact, due to noisy fluctuations. For such cases, the algorithm ignores Peak Velocity 2, since it is very small and does not calculate F_{p3} and the other parameters for the rebound burst. For normal saccades, F_{p3} is assumed to be equal to F_{ts} and T_4 is set to a large value.

1.4.5   CORRECTIONS

Some corrections are made at the end of the initial estimation algorithm. The agonist pulse duration, T_2, and the antagonist onset delay, $T_3 - T_2$, are estimated again if the model-generated saccade does not fit the target displacement. The algorithm increases T_2 and $T_3 - T_2$ into an acceptable range if the estimated final saccade displacement is smaller than the final displacement of data, and decreases T_2 and $T_3 - T_2$ if it is the contrary. Statistically, the correction of increasing T_2 and $T_3 - T_2$ mainly occurs in the case of large saccades due to their large displacement.

If T_2 and $T_3 - T_2$ are adjusted, the parameters for antagonist rebound burst are estimated again. The rebound burst duration is constrained between 3 ms and 14 ms, according to physiological observations (Enderle, 2002). Finally, the start time of the saccade is corrected to a small range around the previous estimate to achieve the smallest value of the objective function. This step is necessary because it is important to match the dynamics of the model to those of the data naturally or some abnormities occur.

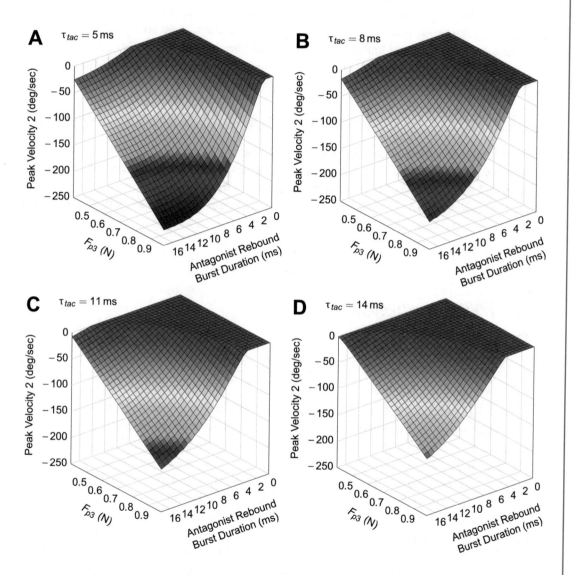

Figure 1.15: The effect of the antagonist rebound burst magnitude F_{p3} and activation time constant τ_{tac} on Peak Velocity 2 under different antagonist rebound burst durations $T_4 - T_3 - 5\tau_{tn2}$.

1.4.6   IMPLEMENTATION

In the implementation of an algorithm, some parameters are transformed based on their physical meanings so that they always have a lower bound (i.e., the algorithm uses $F_{p1} - F_{gs}$ as a parameter instead of F_{p1}, since F_{p1} is always larger than F_{gs}). The initial estimates are loaded into the system identification subroutine and undergo an iteration process using the conjugate gradient method. For most cases studied, the initial guess algorithm demonstrates a good capability in providing a suitable initial estimate. In some cases, the initial guess is very close to the final result.

1.5   INITIAL PARAMETER ESTIMATION FOR MONKEYS

Parameter estimates for monkeys are based on data from the literature. The parameters of interest for monkey extraocular muscles is K_{se}, K_{lt}, B_1, B_2, the empirical equation of the static active-state tension, and the parameters for the oculomotor plant J, B, and K.

The radius of an average rhesus monkey (Macaca mulatta) eyeball is assumed as 10 mm (Fuchs and Luschei, 1971). Thus,

$$\begin{aligned} 1g &= 9.806 \times 10^{-3} \text{ N} \\ 1° &= 1.74 \times 10^{-4} \text{ m.} \end{aligned} \tag{1.68}$$

1.5.1   STATIC CONDITIONS

In the static condition, muscle tension is given by

$$T = \frac{K_{se}}{K_{se} + K_{lt}} F - \frac{K_{se} K_{lt}}{K_{se} + K_{lt}} x_1 \tag{1.69}$$

(see Eq. 5.2.1, Part 1, for details). From the length-tension data for rhesus monkeys, we assume that the length-tension data of rhesus monkeys are straight parallel lines in the operating region of the muscle, with the slope estimated as 0.85 g/° = 47.90 N/m. Assuming that $K_{se} = 125$ N/m in the operating region of monkey extraocular muscles, with the slope of the length-tension curves 47.9 N/m, and the slope given in Eq. (1.69), $\frac{K_{se} K_{lt}}{K_{se} + K_{lt}}$, K_{lt} is estimated as 77.66 N/m. Fuchs and Luschei (1971) suggested that the extraocular muscle tension of monkeys saturates with frequencies above rates of $400° s^{-1}$, which is the upper curve in their Fig. 4. Thus, we assume this curve corresponds to the innervation level of 30°N. At the primary position, the tension is 44 g for 30° N. The equilibrium point of the 2009 linear muscle model is estimated as 25°, according to the passive elasticity, and the muscle length x_1 at the primary position is 4.35 mm (human muscle is 3.705 mm). The empirical equation of the static active-state tension is estimated to match the monkey length-tension curves, and is

$$F = \begin{cases} 0.55 + 0.0175 \, |\theta|, & \text{for } \theta > 0 \\ 0.55 - 0.0125 \, |\theta|, & \text{for } \theta \leq 0. \end{cases}$$

The distribution of length-tension curves from the monkey data is unknown, thus, we assume that the distribution is similar to humans, as shown in Fig. 1.16. Comparing human static active-state tensions to monkey static active-state tensions, we see that the monkey's are generally the larger, corresponding to the higher firing frequency of motoneurons.

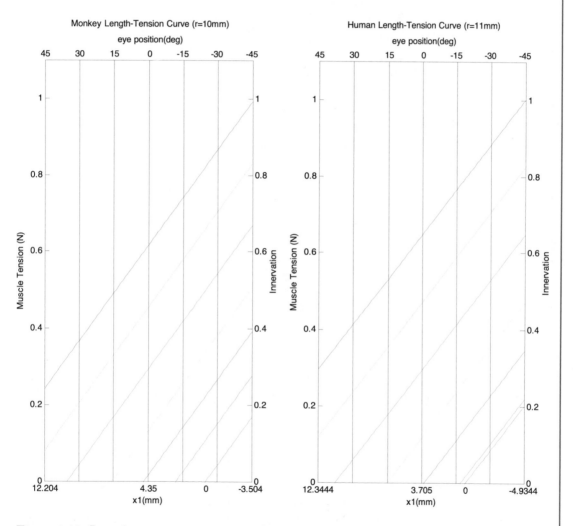

Figure 1.16: Length-tension curves generated by the linear model for monkey extraocular muscles (left). The innervations from upper to lower are 45N, 30N, 15N, 0, 15T, 30T, and 45T (passive). On the right are length-tension curves of human extraocular muscles.

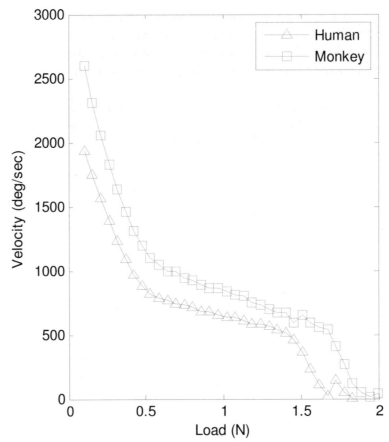

Figure 1.17: Force-velocity relationships for the model of monkeys and humans. (Human: $B_1 = 5.6\,\mathrm{Ns/m}$ and $B_2 = 0.5\,\mathrm{Ns/m}$, Monkey: $B_1 = 4\,\mathrm{Ns/m}$ and $B_2 = 0.4\,\mathrm{Ns/m}$.)

1.5.2   FORCE-VELOCITY CHARACTERISTICS

Parameters B_1 and B_2 describe the force-velocity characteristics of extraocular muscles. Widrick et al. (1997) compared the contractile properties of rat, rhesus monkey, and human type I muscle fibers and determined the force-velocity relationships of rat, monkey, and human soleus type I fiber. Based on this study, we estimate that the monkey extraocular muscles contract faster than human extraocular muscles. Thus, B_1 of the monkey is smaller than B_1 of the human.

Since B_1 of the human is 5.6 Ns/m, B_1 of the monkey is estimated as 4 Ns/m, since the peak velocity of human muscle is around 70% of the peak velocity of the monkey. B_2 of the monkey is estimated as 0.4 Ns/m. The linear muscle model's force-velocity relationships are shown in Fig. 1.17.

1.5.3   OCULOMOTOR PLANT PARAMETERS

K is first determined by steady-state analysis of the model,

$$\frac{180}{\pi r} K_{se}(F_{g0} - F_{t0}) = P_0 \theta$$
$$= (K_{st} K + 2 K_{lt} K_{se}) \theta. \tag{1.70}$$

K should satisfy the steady-state of all eye positions, which yields a value of 10.21 N/m. B is then determined by assuming the dominant orbital time constant is 0.02 s, and

$$\frac{B}{K} = 0.02 \text{ s}, \tag{1.71}$$

which yields $B = 0.204$ Ns/m.

J is the moment of inertia of the eyeball. Since the radius of a monkey eye ball is smaller than the radius of a human eye ball, and assuming that the densities are similar, the moment of inertia of the monkey eye ball is estimated as 80% of the human eye ball, $J = 1.76 \times 10^{-3}$ Ns^2/m.

A comparison of the parameters estimated for monkeys and the parameters of humans is detailed in Table 1.2.

Table 1.2: Comparison of parameters for monkey and human

Parameter	Human	Rhesus Monkey
Radius of eye ball	11 mm (11.8 mm in model)	10 mm
K_{se}	125 N/m	125 N/m
K_{lt}	60.7 N/m	77.66 N/m
B_1	5.6 N s/m	4 N s/m
B_2	0.5 N s/m	0.4 N s/m
F	$\begin{cases} 0.4 + 0.0175\lvert\theta\rvert, \text{ for } \theta > 0 \\ 0.4 - 0.0125\lvert\theta\rvert, \text{ for } \theta \leq 0 \end{cases}$	$\begin{cases} 0.55 + 0.0175\lvert\theta\rvert, \text{ for } \theta > 0 \\ 0.55 - 0.0125\lvert\theta\rvert, \text{ for } \theta \leq 0 \end{cases}$
K	16.34 N/m	10.21 N/m
B	0.327 N s/m	0.204 N s/m
J	2.2×10^{-3} N s^2/m	1.76×10^{-3} N s^2/m

Compared to the human muscle, the monkey muscle is stiffer, reflected by the larger elastic parameter K_{lt}. It also contracts faster than human muscle, since it has smaller B_1. The firing frequency of monkey motoneurons is faster than humans. The oculomotor plant parameters of monkey extraocular muscles are smaller than the ones of humans because the monkey eye ball is smaller.

The transfer function for the oculomotor plant, based on Eq. (1.26), is

$$H(s) = \frac{\theta}{\Delta F} = \frac{\delta B_2 \left(s + \dfrac{K_{se}}{B_2} \right)}{s^3 + P_2 s^2 + P_1 s + P_0}, \tag{1.72}$$

where $\Delta F = F_{ag} - F_{ant}$. Using the parameter values in Table 1.2, we have the transfer function for humans as

$$H(s) = \frac{1.9406 \times 10^5 (s + 250)}{s^3 + 596 s^2 + 1.208 \times 10^5 s + 1.3569 \times 10^6} \tag{1.73}$$

and for monkeys

$$H(s) = \frac{2.6904 \times 10^5 (s + 312.5)}{s^3 + 575.2 s^2 + 1.4829 \times 10^5 s + 2.7743 \times 10^6}. \tag{1.74}$$

There are three poles and the one zero in the transfer function shown in Eq. (1.72). Using the parameter values in Table 1.2 for humans, the poles are

$$-292.22 + j168.63$$
$$-292.22 - j168.63$$
$$-11.92$$

and the zero is

$$250.$$

For monkey, the poles are

$$-277.48 + j245.09$$
$$-277.48 - j245.09$$
$$-20.24$$

and the zero is

$$312.5.$$

For humans, the time constant for the real pole is 3.4 ms, and for the complex pole, 83.9 ms. Similarly, for monkeys we have 3.6 ms and 49.4 ms.

Data and analysis of the monkey data is presented first in the next section. Following this, data and analysis of human data is presented, with inferences on neural control signals based on the monkey.

1.6   MONKEY DATA AND RESULTS

Data[3] were collected from a rhesus monkey that executed a total of 27 saccades in our data set for 4°, 8°, 16°, and 20° target movements. Neuron data were recorded from the long lead burst

[3]Details of the experiment and training are reported elsewhere (Sparks et al., 1976; data provided personally by Dr. David Sparks).

neuron (5 saccades), excitatory burst neuron (17 saccades), and the agonist burst-tonic neuron (5 saccades). The firing of the burst tonic neuron is similar to the motoneuron that drives the agonist muscle during a saccade. Figure 1.18 shows the estimation results for three saccades (4°, 8°, and 15°). Initial estimates of monkey model parameters are listed in Table 1.2. The system identification technique provides close agreement between the displacement and velocity estimates from the model, and the displacement data and the derivative of the displacement data (velocity). The acceleration estimate from the model was the least accurate when compared with the second derivative of the displacement data (acceleration); it should be noted that the second derivative of the displacement data considerably amplifies the noise in the data. The accuracy of the results in Fig. 1.18 is typical for the other 4°, 8°, 16°, and 20° saccadic eye movements in our study.

Figure 1.19 shows the estimated neural inputs and active-state tensions that generate the saccades shown in Fig. 1.18. Also shown are the firing rates recorded from a single burst-tonic cell in a rhesus monkey (green line) for these saccades, scaled to match the height of N_{ag}. The shapes of the model's neural inputs approximates the burst-tonic data during the pulse and slide very closely. The estimated agonist neural input N_{ag} clearly has a similar shape as the firing rate data. It should be noted that the firing activity in the data comes from a single burst-tonic neuron. The neural input to the oculomotor plant is actually due to the firing of more than 1,000 motoneurons.

To evaluate the robustness of the estimation results, the neuron data are filtered by a discrete low-pass filter to obtain neuron-data-derived active-state tensions using Eq. (1.58). The filter uses the activation and deactivation muscle time constants estimated by the system identification previously described. A firing threshold is set to neglect the random and small firing before the bursting of the neuron. Figure 1.20 shows the data-derived active-state tensions. The neuron-data-derived active-state tensions are very close to those estimated by the system identification technique shown in Fig. 1.20. Using the data-derived active-state tensions as inputs to the 2009 oculomotor plant model, Fig. 1.20 also shows the generated eye position, velocity, and acceleration plotted against the data.

As with the results presented in Fig. 1.18, the data is closely approximated by the predictions of the model. It should be noted that no post-saccade phenomena (i.e., those with a dynamic overshoot or a glissade) are observed in any of the monkey data analyzed, and no PIRB in the antagonist muscle was detected by the system identification technique.

Important parameter estimates for all 27 saccades are shown in Figs. 1.21 and 1.22. The main-sequence diagram is shown in Fig. 1.21. Peak-velocity estimates follow an exponential shape as a function of saccade magnitude. Duration has a linear relationship with saccade magnitude for saccades above approximately 8°. For saccades less than 8°, duration is approximately constant with high variability. It should be noted that saccade duration is difficult to determine, especially for small saccades, and may be a source of differences with other published data. The latent period is relatively independent of saccade magnitude.

The estimated agonist pulse magnitudes and durations are shown in Fig. 1.22 for all 27 saccades. The agonist pulse magnitude does not significantly increase as a function of saccade magni-

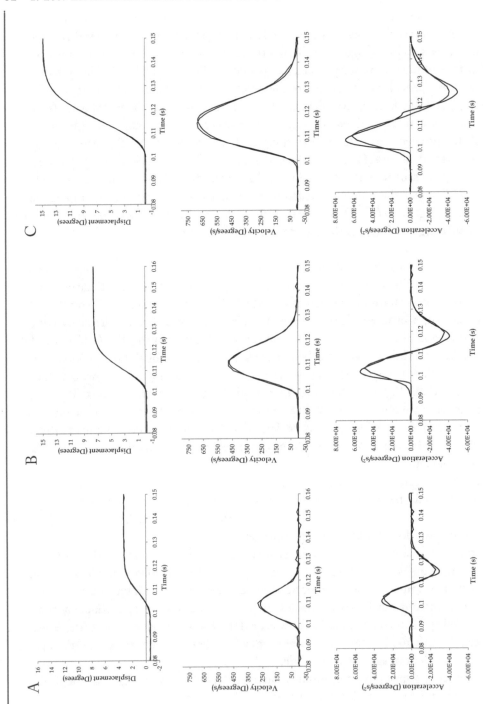

Figure 1.18: Eye position velocity and acceleration for three different saccades in a rhesus monkey (A: 4°, B: 8°, and C: 15°). Blue lines are the model predictions and the red lines are the experimental data during the saccadic eye movement.

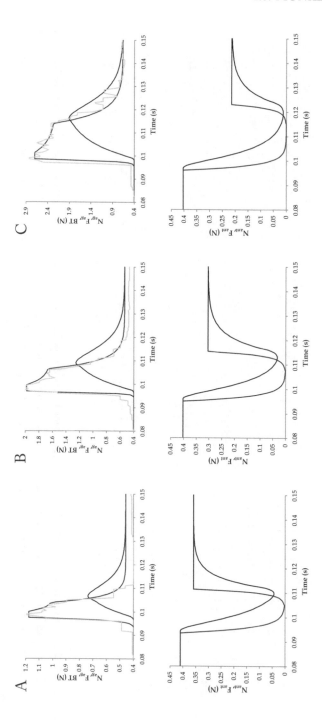

Figure 1.19: The estimated agonist and antagonist neural inputs N_{ag} and N_{ant} (red line), active–state tension F_{ag} and F_{ant} (blue line) for the three saccades (A: 4°, B: 8°, and C: 15°) shown in Fig. 1.18. Also shown are the firing rates recorded from a single burst-tonic cell in a rhesus monkey (green line) for these saccades, scaled to match the height of N_{ag}.

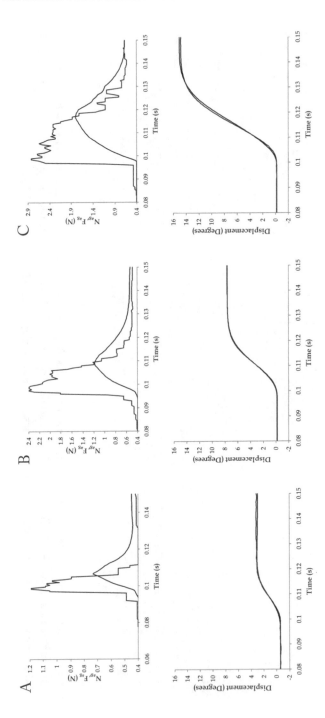

Figure 1.20: Active-state tensions (blue line) obtained from low-pass filtered firing rate of a burst-tonic neuron (red line) using the activation and deactivation time constants calculated in the parameter estimation. Also shown are the model predictions using the parameter estimates from the system identification technique for displacement, velocity and acceleration, and the data (red line is data and blue line is the model predictions). (A: 4°, B: 8°, and C: 15°).

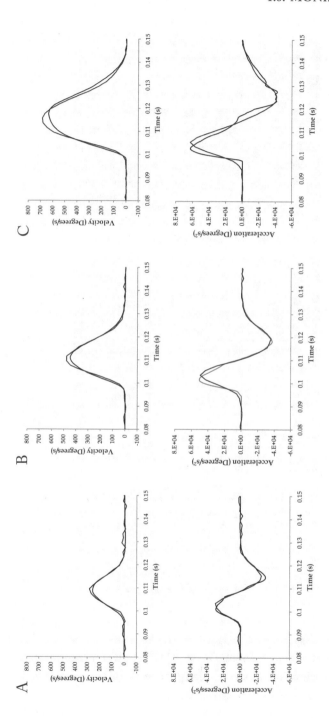

Figure 1.20: *(Continued)*. Active-state tensions (blue line) obtained from low-pass filtered firing rate of a burst-tonic neuron (red line) using the activation and deactivation time constants calculated in the parameter estimation. Also shown are the model predictions using the parameter estimates from the system identification technique for displacement, velocity and acceleration, and the data (red line is data and blue line is the model predictions). (A: 4°, B: 8°, and C: 15°).

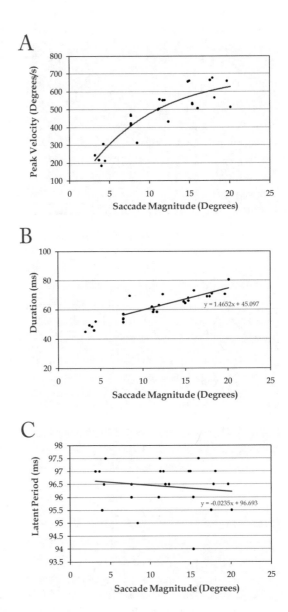

Figure 1.21: Main sequence diagram for monkey data. (A) Peak velocity vs. saccade magnitude. Fitted exponential curve to the data is $\dot{\theta}_{pv} = 694.65(1 - e^{-0.1155\theta_{ss}})$. (B) Duration vs. saccade magnitude with regression straight approximation for saccades larger than 5°. (C) Latent period vs. saccade magnitude, based on the data.

tude for saccades larger than 8°, consistent with the time-optimal controller proposed by Enderle (2002). For saccades under 8°, agonist pulse magnitude shows a linear increase in pulse magnitude vs. saccade magnitude, again in agreement with our theory for the saccade controller. A great variability is observed in the pulse magnitude estimates for saccades of the same magnitude, which is also observed by Hu et al. (2007) in their analysis of the firing rates in the monkey EBN. The agonist pulse duration increases as a function of saccade magnitude for saccades larger than 8°, and for smaller saccades, the duration of the agonist pulse is relatively constant as a function of saccade magnitude. Note that for all saccades, the pulse magnitude is tightly coordinated with the pulse duration. Interestingly, Fuchs and coworkers describe the firing rate for saccades greater than 10° as saturating, and for saccades up to 5°, they have the same approximate duration in monkeys (Fuchs et al., 1985a). We will discuss these details when the time optimal controller is presented.

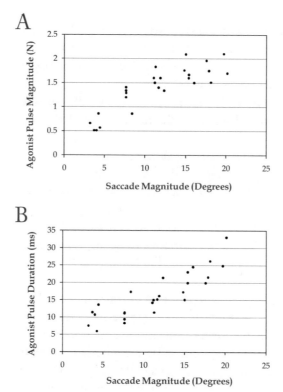

Figure 1.22: Agonist pulse magnitude (A) and duration (B) as functions of saccade magnitude for monkey data.

1.7   HUMAN DATA AND RESULTS

Data[4] was collected from three human subjects executing 127 saccades, many with dynamic overshoots or glissades. Displayed in Fig. 1.23 are representative model estimates of saccades generated with a dynamic overshoot, a glissadic overshoot, and normal characteristics. The model predictions for all saccades match displacement data and estimates of velocity very well, including saccades with a dynamic or a glissadic overshoot, with accuracy similar to those in Fig. 1.23. As noted before, the acceleration model prediction and the data-derived acceleration estimate were not as good as those of displacement and velocity.

The 8° saccade shown in Fig. 1.23A of data and model predictions has dynamic overshoot. Note that the saccade with dynamic overshoot is caused by a PIRB firing in the antagonist neural input at approximately 220 ms. The PIRB induces prominent reverse peak velocity as shown.

Figure 1.23B shows model predictions and data for a 8° saccade with glissadic overshoot. The glissade is caused by the PIRB in the antagonist neural input at approximately 223 ms. Notice the peak firing for a saccade with glissadic overshoot is smaller than one with dynamic overshoot. The PIRB induces reverse peak velocity that is smaller than the one with dynamic overshoot. In glissadic overshoot, the eye has an overshoot that returns to steady-state more gradually. As a result, the glissade has a smaller peak velocity.

A −12° normal saccade is shown in Fig. 1.23C. Normal saccades usually do not have a PIRB, although this is not absolute, as the timing of the PIRB might offset the impact of the burst.

Important parameter estimates for all 127 saccades are shown in Figs. 1.24–1.27. The main-sequence diagram is shown in Fig. 1.24. Peak-velocity estimates from the model are in close agreement with the data estimates of peak velocity and follow an exponential shape as a function of saccade magnitude. Duration has a linear relationship with saccade magnitude for saccades above 7°. For saccades between 3° to 7°, duration is approximately constant. It should be noted that saccade duration is difficult to determine, especially for small saccades, and may be a source of differences with other published data. The latent period is relatively independent of saccade magnitude.

The estimated agonist pulse magnitudes and durations are shown in Fig. 1.25 for all 127 saccades. The agonist pulse magnitude does not significantly increase as a function of saccade magnitude for saccades larger than 7°, consistent with the time-optimal controller proposed by Enderle (2002). For saccades under 7°, agonist pulse magnitude shows a linear increase in pulse magnitude vs. saccade magnitude, again in agreement with our theory for the saccade controller. A great variability is observed in the pulse magnitude estimates for saccades of the same magnitude, which is also observed by Hu et al. (2007) in their analysis of the firing rates in the monkey EBN. The agonist pulse duration increases as a function of saccade magnitude for saccades larger than 7°. For saccades between 3° to 7°, the duration of the agonist pulse is relatively constant as a func-

---

[4]Details of the experiment are reported in Enderle and Wolfe (1988a).

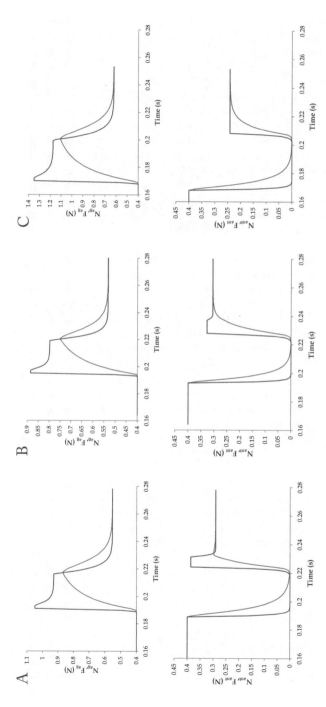

Figure 1.23: (A) Dynamic overshoot saccade of 8°; (B) glissadic overshoot saccade of 8°; and (C) normal −12° saccade. The first two lines of the graphs are the active-state tension and neural input calculated from the parameter estimation. Also shown are the model predictions using the parameter estimates from the system identification technique for displacement, velocity and acceleration, and the data (red line is data and blue line is the model predictions).

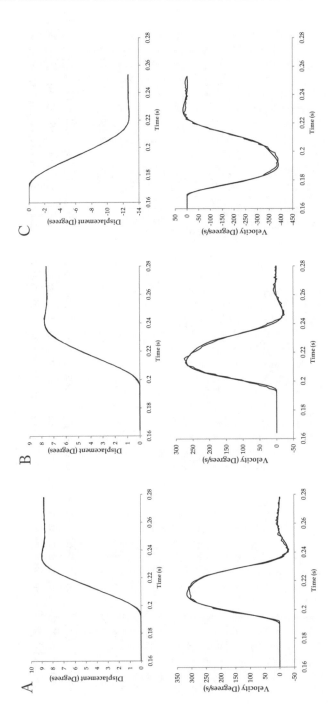

Figure 1.23: *(Continued.)* (A) Dynamic overshoot saccade of 8°; (B) glissadic overshoot saccade of 8°; and (C) normal −12° saccade. The first two lines of the graphs are the active-state tension and neural input calculated from the parameter estimation. Also shown are the model predictions using the parameter estimates from the system identification technique for displacement, velocity and acceleration, and the data (red line is data and blue line is the model predictions).

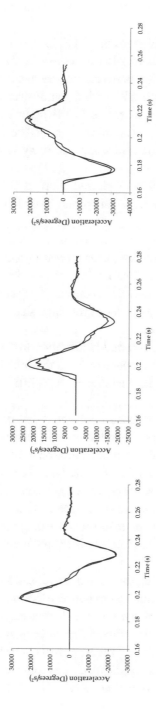
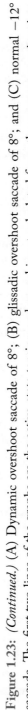

Figure 1.23: *(Continued.)* (A) Dynamic overshoot saccade of 8°; (B) glissadic overshoot saccade of 8°; and (C) normal −12° saccade. The first two lines of the graphs are the active-state tension and neural input calculated from the parameter estimation. Also shown are the model predictions using the parameter estimates from the system identification technique for displacement, velocity and acceleration, and the data (red line is data and blue line is the model predictions).

tion of saccade magnitude. Note that for all saccades, the pulse magnitude is tightly coordinated with the pulse duration.

Figures 1.26 and 1.27 provide estimates for the PIRB in the antagonist motoneuron, where the PIRB induces a reverse peak velocity. Normal saccades usually do not have a notable rebound burst. Figure 1.26 describes the relationship between saccade magnitude and PIRB magnitude and duration. As shown, there are more saccades with dynamic overshoot in the abducting than adducting direction. There is also great randomness in the occurrence of PIRB. As shown in Fig. 1.26, the rebound burst magnitudes for dynamic overshoots are usually larger than the magnitudes for glissades. The rebound burst duration is approximately 12 ms, with considerable variation for saccades of the same size.

In Fig. 1.27, the antagonist onset delay is plotted. Normal saccades are clustered close to the origin, while moving further from the origin, glissades cluster in a band followed by saccades with a dynamic overshoot.

Shown in Fig. 1.28 are the estimates for the agonist and antagonist time constants. The agonist and antagonist activation time constants have a great impact on saccade dynamics. The agonist activation time constant showed great variability for saccades of the same size. No trends were observed between the agonist activation time constant and saccade magnitude. The other time constants showed less variability for saccades of the same size. Other parameter values had little variation around their initial estimates.

Figure 1.29 shows the results of the system identification for estimating the parameters of the oculomotor plant. B_1 has the greatest movement from the initial parameter estimate, especially for positive eye movements. The other parameters have little variation about their initial values.

The objective of this chapter is to present a new model that more accurately characterizes horizontal saccadic eye movements using a 3rd-order linear homeomorphic model of the oculomotor plant with a pulse-slide-step neural controller. Also included in the controller is PIRB in both the agonist and antagonist neural controller after marked inhibition. Of fundamental importance is the oculomotor plant used in the study of the saccade controller. If the oculomotor plant is not homeomorphic, then the controller obtained is suspect. Our model is robust in accurately simulating horizontal saccades that follow the main-sequence diagram, and a linear muscle model that is homeomorphic and has the nonlinear force-velocity and length-tension characteristics of muscle.

Model parameters are estimated using the system identification technique with both monkey and human data showing an excellent agreement between the model predictions and the data. Analysis of the monkey data gives support that we are able to accurately estimate agonist neural input during saccades, and, by inference, to accurately estimate agonist neural input in humans. Another purpose of this chapter is to describe a time-optimal controller for horizontal saccades. Although discussed more completely in Chapter 2 we discuss some features of the neural network here in the context of PIRB and the time-optimal controller. It is thought that the EBN, located

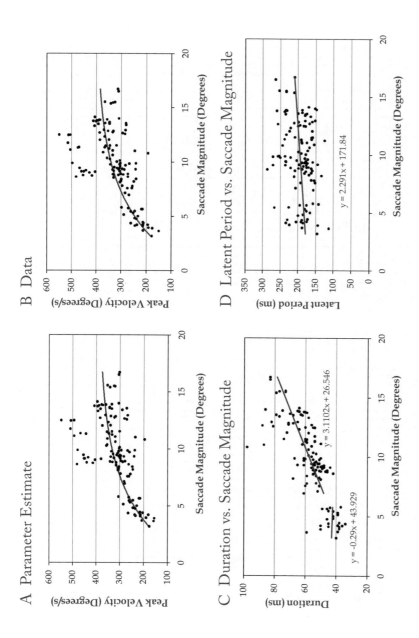

Figure 1.24: Main-sequence diagram for all 127 saccades from three human subjects. (A) Peak velocity vs. saccade magnitude from the model estimates, with regression fit $\dot{\theta}_{pv} = 390(1 - e^{-0.2\theta_{ss}})$. (B) Peak velocity vs. saccade magnitude from the data, with regression fit $\dot{\theta}_{pv} = 401(1 - e^{-0.2\theta_{ss}})$. (C) Duration vs. saccade magnitude based on the data. (D) Latent period vs. saccade magnitude. Note that the parameter estimation program did not update the duration or the latent period, thus a single graph for each is drawn.

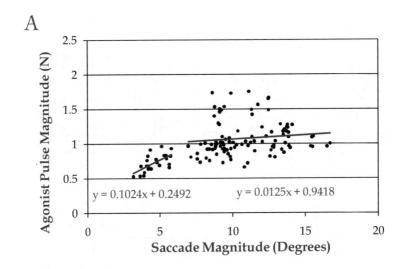

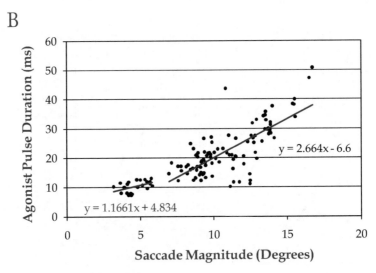

Figure 1.25: Agonist pulse magnitude (A) and duration (B) as functions of saccade magnitude.

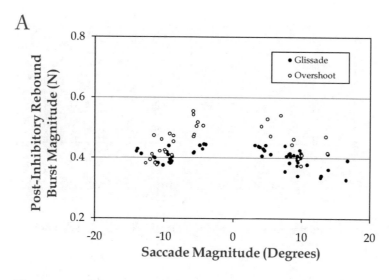

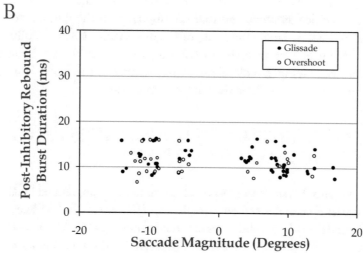

Figure 1.26: Post-inhibitory rebound burst magnitude (A) and duration (B) as functions of saccade magnitude.

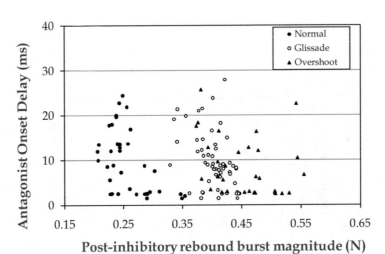

Figure 1.27: Post-saccade phenomena involving normal, glissade, and dynamic overshoot saccades.

in the paramedian reticular formation (PPRF), monosynaptically projects to motoneurons (either the abducens or oculomotor), and that the discharge in the motoneuron during a saccade resembles a delayed EBN signal. The firing of the motoneurons is responsible for movement of the eyes. Another burst neuron in the PPRF, called the inhibitionary burst neuron (IBN), is also active during saccades with a discharge pattern similar to the EBN. IBNs are involved with inhibiting the neuron sites that drive the antagonist muscle.

1.8   POST-INHIBITORY REBOUND BURST AND POST-SACCADE PHENOMENA

Inhibition of antagonist burst neurons is postulated to cause an unplanned PIRB toward the end of a saccade that causes dynamic overshoots or glissades (Enderle, 2002). While some studies do not observe the rebound firing in the abducens neurons in monkeys (Fuchs and Luschei, 1970, Ling et al., 2007, Sylvestre and Cullen, 2006), PIRB are observed in the abducens motoneurons at the end of off-saccades in monkeys in other studies (for examples, see Robinson (1981), Van Gisbergen et al. (1981)). It has been noted earlier that saccades with dynamic overshoots or glissades do not occur with the same frequency in the monkey as in humans, and that they are absent from our monkey data.

Our theory is that, at least in humans, the antagonist PIRB causes a reverse peak velocity during dynamic overshoots or glissades in humans. The model predictions accurately match the velocity data for the entire saccade, including saccades with dynamic or glissadic overshoot. We

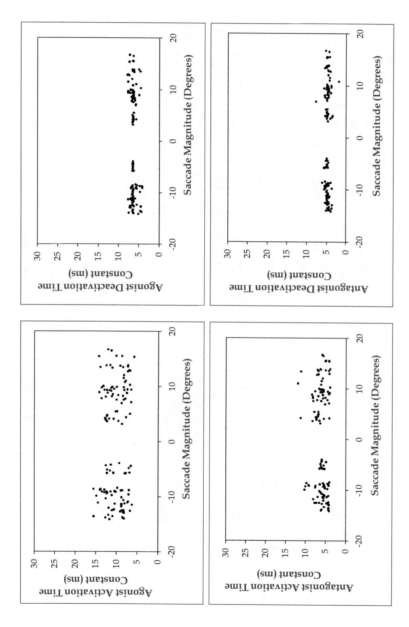

Figure 1.28: Activation and deactivation time constants for the agonist and antagonist muscles.

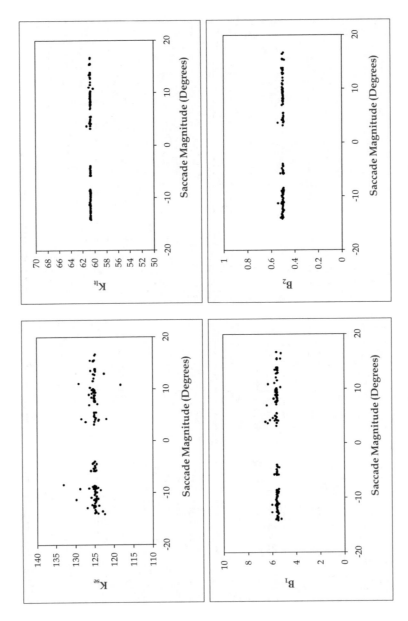

Figure 1.29:   Final parameter estimates for the oculomotor plant obtained from 127 saccades.

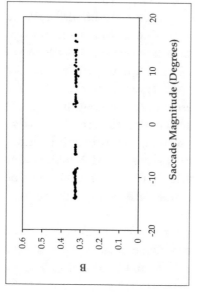

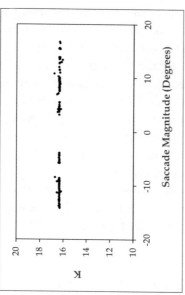

Figure 1.29: *(Continued.)* Final parameter estimates for the oculomotor plant obtained from 127 saccades.

were unable to generate saccades with post-saccade behavior based on just the timing of the antagonist step, but we needed the PIRB to generate saccades with dynamic or glissadic overshoot.

Figures 1.26 and 1.27 summarize the characteristics of the 127 saccades collected from the three human subjects. The number of saccades with a glissade is larger than the number of normal saccades or those with a dynamic overshoot. Additionally, the incidence of dynamic overshoot decreases as saccade size increases. As shown, saccades with a dynamic overshoot typically have larger rebound burst magnitude than those with a glissade or with normal characteristics. The antagonist onset delay varies from 3 ms to approximately 25 ms. With a larger rebound burst, the onset delay is typically shorter for each type of saccade.

An inherent coordination error exists between the return to tonic firing levels in the abducens and oculomotor motoneurons during the completion of a saccade. During an abducting saccade, ipsilateral abducens motoneurons fire without inhibition and oculomotor motoneurons are inhibited during the pulse phase. Because the IBN inhibits antagonist motoneurons, resumption of tonic firing and PIRB activity in the motoneurons does not begin until shortly after the ipsilateral IBNs cease firing. This same delay exists in the abducens motoneurons for adducting saccades.

There are significantly more internuclear neurons between the contralateral EBN and the TN and the ipsilateral oculomotor motoneurons (antagonist neurons during an abducting saccade), than the ipsilateral EBN and TN and ipsilateral abducens motoneurons (antagonist neurons during an adducting saccade). Due to the greater number of internuclear neurons operating during an abducting saccade, a longer time delay exists before the resumption of activity in the oculomotor motoneurons after the pulse phase for abducting than adducting saccades.

Since the time delay before the resumption of activity in the oculomotor motoneurons after the pulse phase of a saccade is greater for abducting saccades than with adducting saccades, the incidence of saccades with dynamic overshoot should be greater for abducting saccades than adducting saccades. This is precisely what is observed in saccadic eye movement recordings; most saccades with dynamic overshoot occur in the abducting direction. Additionally, because the contralateral TN's firing rate decreases as ipsilateral saccade amplitude increases, the rate of dynamic overshoot decreases, since fewer saccades have sufficiently high PIRB magnitudes. This is also what is observed in saccadic eye movement recordings.

Figure 1.30 illustrates the relationship between the PIRB magnitude and the antagonist onset time delay. As shown, it is possible for a normal saccade to have a small PIRB, as long as the onset delay is small. As the onset delay increases, the PIRB must decrease or a saccade with dynamic or glissadic overshoot occurs.

1.9   TIME-OPTIMAL CONTROLLER

The general principle for a time-optimal controller for the horizontal saccade system is that the eyes reach their destination in minimum time that involves over 1,000 neurons. Each neuron contributes to the neural input to the oculomotor plant. In Section 3.6, Part 1, we described the

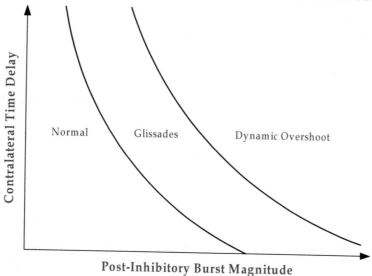

Figure 1.30: Relationship between antagonist onset time delay and the post-inhibitory rebound burst magnitude.

time-optimal control of saccadic eye movements with a single switch-time using a linear homeomorphic oculomotor plant for the lateral and medial rectus muscles. This section reexamines this earlier study using the updated oculomotor plant and a time-optimal controller constrained by a more realistic pulse-slide-step motoneuron stimulation of the agonist muscle with a pause and step in the motoneuron stimulation of the antagonist muscle, and physiological constraints.

The time-optimal controller proposed here has a firing rate in *individual* neurons that is maximal during the agonist pulse and independent of eye orientation, while the antagonist muscle is inhibited. We refer to maximal firing in the neuron as the intent of the system, which, because of biophysical properties of the neuron membrane, slowly decay over time, as described in Chapter 2. The type of time-optimal controller described here is more complex than the one previously described in Chapter 3, Part 1, due to physiological considerations in the SC. The time-optimal controller operates in two modes, one for small saccades and one for large saccades.

The duration of small saccades has been reported as approximately constant (Fuchs et al. (1985a) and reported here), and also as a function of saccade amplitude (e.g., Bahill et al. (1980)). Estimating the saccade start and end time is quite difficult because it is contaminated by noise. We used a Kaiser filter to reduce the impact of noise, which others may not have implemented, and possibly introduced a difference in results. Moreover, synchrony of firing will have a greater impact on the start time for small saccades than larger saccades, since the beginning of the saccades is much more drawn out, making detection more difficult. In our analysis, a regression fit for the data is carried out in two intervals, one between 3° and 7°, and one for those greater than 7°.

Our results indicate an approximately constant duration for small saccades and a duration that increases with saccade size for large saccades. Other investigators have used a single interval for the regression fit to a straight line, or a nonlinear function. It is possible that using the technique used here will result in a similar conclusion to ours. Since we did not analyze saccades less than $3°$, judgment on saccade duration in this interval is delayed and supports future investigation.

We propose that there is a minimum time period that EBNs can be switched on and off, and that this is a physical constraint of the system. As shown in Fig. 1.24C, small saccades have approximately the same duration of 44 ms that does not significantly change as a function of saccade magnitude. Also, note that there is randomness in the response, where saccades with large pulse magnitudes are matched with shorter durations, and vice versa. As the saccade size increases for small saccades, we propose that additional neurons be added to the agonist neural input up to $7°$, where, above this, all neurons are engaged.

In our model, we sum the input of all active motoneurons into the firing of a single neuron. Thus, as the magnitude of the saccades increases, the firing rate of the single neuron in our model increases up to $7°$, after which it is maximal, since all neurons are firing. Keep in mind, however, that the firing rate of a real neuron is maximal and does not change as a function of saccade magnitude, as easily seen in Fig. 4 in Robinson (1981) and Fig. 2 in Van Gisbergen et al. (1981). The overall neural input for the agonist pulse is given by

$$N_{ag} = \begin{cases} N(\theta_T)N_{ag_i} & \theta < 7° \\ N_{ag_{max}} & \theta \geq 7°, \end{cases} \tag{1.75}$$

where $N(\theta_T)$ is the number of neurons firing for a saccade of θ_T degrees, N_{ag_i} is the contribution from an individual neuron, and $N_{ag_{max}}$ is the combined input from all neurons. For small saccades, the commencement of firing of the individual neurons, or synchrony of firing, has a great impact on the overall neural input, since the period of firing during the pulse is small (10 ms for the estimate in Fig. 1.25B). Randomness in the start time among the active neurons means that the beginning of the saccade is more drawn out than if they all started together. For smaller saccades, this may result in an incorrect start time, which then effects the duration. Any lack of synchrony can cause the overall agonist input to be smaller; this is a much larger factor for a small saccade than a large saccade, since the pulse duration is much larger. It is very likely that, during a saccade, neurons do not all commence firing at the same instant. This is seen in Fig. 1.25B where there is a small slope to the regression fit.

Above $7°$, the magnitude of the saccade is dependent on the duration of the agonist pulse with all neurons firing maximally. The agonist pulse magnitude as shown in Fig. 1.25A is approximately a constant, according to the regression fit. The duration of the agonist pulse increases as a function of saccade magnitude, as shown in Fig. 1.25B.

The saccade controller described here is a time-optimal controller that differs from the one describe by Enderle and Wolfe (1987) because of the physiology of the system. Active neurons during the pulse phase of the saccade all fire maximally. For saccades greater than $7°$, this is the same time-optimal controller described earlier by Enderle and Wolfe (1987). For saccades from 3° to 7°, the system is constrained by a minimum duration of the agonist pulse; saccade magnitude is dependent on the number of active neurons, all firing maximally, consistent with physiological evidence. In terms of control, it is far easier to operate the system for small saccades based on the number of active neurons firing maximally, rather than adjusting the firing rate for all neurons as a function of saccade magnitude as proposed by others. Thus, the system described here is still time-optimal based on physiological constraints.

Recently, Harris and Wolpert (2006) described a saccade controller that optimizes speed and accuracy in support of a time-optimal controller, the same type of controller used by Enderle and Wolfe (1987). Harris and Wolpert's 3-pole oculomotor plant was not homeomorphic as the plant described here, nor was their controller based on neuro-anatomical constraints or able to produce realistic saccades, whether normal, or containing dynamic overshoots or glissades. Their numerator term in the oculomotor plant did not include a derivative term as suggested by our model in Eq. (1.26), which has a significant impact on the results. The main sequence diagram presented in their Fig. 2 does not have the characteristics of those reported elsewhere, such as Bahill et al. (1980), with a leveling off of peak velocity at approximately 20°. This could be due to the oculomotor plant used in their model.

Generally, saccades recorded for any size magnitude are extremely variable, with wide variations in the latent period, time to peak velocity, peak velocity, and duration.

Furthermore, this variability is well coordinated by the neural controller. Saccades with lower peak velocity are matched with longer saccade durations, and saccades with higher peak velocity are matched with shorter saccade durations. Thus, saccades driven to the same destination usually have different trajectories.

Hu et al. (2007) examined the variability in saccade amplitude, duration, and velocity in the monkey by recording eye position and the EBN. To examine the reliability of the EBN, saccades with the similar amplitude and velocity were analyzed, and it was determined that the initial portion of the EBN firing rate had little variability, while the last portion of the burst had observable variability. The initial portions of the burst for a 10° and 20° saccade shown in their Fig. 2 are approximately the same size and shape. The major difference between the 10° and 20° saccade is that the 20° saccade had a longer burst duration. Furthermore, Hu and coworkers proposed that the activity in a single burst cell is not independent of, but strongly correlated with the activity of other burst neurons. To achieve the low variability in the EBN burst for the population, a low variability in the input or a special biophysical property of the burst neurons exists, or a combination of the two is proposed by Hu et al. (2007).

CHAPTER 2

A Neuron-Based Time-Optimal Controller of Horizontal Saccadic Eye Movements and Glissades

2.1 INTRODUCTION

The control mechanism[1] of the human binocular vision is staggering in its complexity, and has stunned many neuroscientists in their quest to match its functionality to a greater or lesser degree. As described in Chapter 1, saccades are categorized into two different modes of operation: small (below 7°) and large (above 7°). The differentiation between these two modes is based on the fact that, when the saccade size increases, more active motoneurons are firing synchronously to form the agonist neural input for small saccades. For large saccades, however, the number of active motoneurons firing maximally remains unchanged, and the agonist pulse duration is directly related to the saccade magnitude.

The saccade neural network requires involvement of a series of neurons designed to imitate the behavior of actual neuronal populations in the horizontal saccade controller. Considerable research has been concerned with developing a generic neuron model to provide the means to develop a network of neurons, tailored to the complexity involved with inherent physiological evidence. The widespread utility of spiking neural networks (SNNs) lies in leveraging efficient learning algorithms to the spike response models (Ghosh-Dastidar and Adeli, 2007, 2009). A spike-pattern association neuron identified five classes of spike patterns associated with networks of 200, 400, and 600 synapses, with success rates of 96%, 94%, and 90%, respectively (Mohemmed et al., 2012). Rossell'o et al. (2009) synthesized and tested a hybrid analog-digital circuitry to implement an SNN that outputs the postsynaptic potential by integrating the filtered action potentials. A neural system comprised of a persistent firing sensory neuron, a habituating synapse, and a motoneuron was developed to illustrate the spike-timing dependency of the working mem-

---

[1]Some of the material in this chapter is an expansion of previously published papers: Ghahari, Alireza, Zhai, Xiu, and Enderle, J., Development of a Neural Network Model for Controlling Horizontal Saccadic Eye Movements, *Proceedings of the IEEE 39th Northeast Bioengineering Conference*, Syracuse, New York, April 5-7, 2013, pp. 29–30, and Ghahari, A., and Enderle, J.D., A Neuron-Based Time-Optimal Controller of Horizontal Saccadic Eye Movements. *International Journal of Neural Systems*, Vol. 24, 1450017 (19 pages), 2014. DOI: 10.1142/S0129065714500178.

ory (Ramanathan et al., 2012). The persistent firing neuron stems from the Izhikevich neuron model (2003), the habituating synapse is a conductance-based model, and the motoneuron captures the essence of the Hodgkin-Huxley (HH) model (1952). These studies provide an abundance of evidence that an SNN is well suited to evoke the properties of the firing patterns of the premotor neurons during the pulse and slide phases of a saccade. However, none of the studies presented a demonstration of the neural circuits reproducing electrophysiological responses in a network of neurons at both premotor and motor levels. To encompass all of the desired neural behaviors, Enderle and Zhou (2010) proposed a neural circuitry to match the firing rate trajectory of the premotor neurons. We model the saccade-induced spiking activities at the premotor level with an HH model for the bursting neurons, and with a modified FitzHugh-Nagumo (FHN) model for the tonic-spiking neurons (Faghih et al., 2012).

Time-optimal control theory of the horizontal saccade system establishes the fact that there is a minimum time required for the eyes to reach their destination by involving thousands of neurons. As indicated in Section 1.9, conjugate goal-directed horizontal saccades are well characterized by a 1st-order time-optimal neural controller. It is important that this new, more complex time-optimal controller ascertains that the firing rate of the motoneurons does not change as a function of saccade magnitude during the pulse phase. This time-optimal neural controller is consistent with the physiological evidence to generate realistic saccades, whether normal or exhibiting dynamic or glissadic overshoots.

In this chapter, we focus on neural control of horizontal human saccades. A neural network model of saccade-related neural sites in the midbrain is presented first. Next, we characterize the underlying dynamics of each neural site in the network. For this purpose, we emphasize the critical aspects of the neural network dynamics that need to be treated in the case of spiking neurons. In consequence, a saccadic circuitry that includes omnipause neuron (OPN), premotor excitatory burst neuron (EBN), inhibitory burst neuron (IBN), long lead burst neuron (LLBN), tonic neuron (TN), interneuron (IN), abducens nucleus (AN), and oculomotor nucleus (ON) is developed to match the dynamics of the neurons. The computational neural modeling is motivated by discussing the general applicability of SNNs to the biophysical modeling of interconnected neurons. This perspective elucidates broad insights to modeling at different structural scales, such as the circuit and the systems levels, which we develop subsequently. Finally, the motoneuronal control signals drive a time-optimal controller that stimulates the linear homeomorphic model of the oculomotor plant described in Chapter 1. The study of glissades will follow, thus allowing the comparison between them and the normal saccades. The computational modeling of glissades is insightful because it allows investigation of one of the widely reported oculomotor version dysfunctions.

The proposed saccadic circuitry is a complete model of saccade generation since it not only includes the neural circuits at both the premotor and motor stages of the saccade generator, but it also uses a time-optimal controller to yield the desired saccade magnitude. We abbreviate the "conjugate goal-directed horizontal human saccade" with the term "saccade" throughout this chapter. The terms "motoneurons" and "agonist (antagonist) neurons" are also substitutable.

2.2   NEURAL NETWORK

Neurophysiological evidence and developmental studies indicate that important neural populations, consisting of the cerebellum, superior colliculus (SC), thalamus, cortex, and other nuclei in the brainstem, are involved in the initiation and control of saccades (Coubard, 2013, Enderle, 1994, Enderle and Engelken, 1995, Enderle, 2002, Enderle and Zhou, 2010, Zhou et al., 2009). The studies also provided evidence that saccades are generated through a parallel-distributed neural network, as shown in Fig. 2.1. The two sides of the symmetric network in Fig. 2.1 are known as the ipsilateral side and the contralateral side. The ipsilateral side exhibits coordinated activities in the initiation and control of the saccade in the right eye, while the contralateral side simultaneously synapses with the ipsilateral side to generate a saccade in the left eye. Each neuron in the parallel-distributed network fires in response to other neurons to stimulate the final motoneurons on both sides of the network to execute a binocular saccade. The neural populations on each side of the midline excite and inhibit one another sequentially to ensure that this coactivation leads to the coordination of movement between the eyes.

In the context of the neuroanatomical connectivity structure in Fig. 2.1, the saccade neural network includes neuron populations to imitate the behavior of actual neuronal populations in the initiation, control, and termination of the saccadic burst generator. Neural coordinated activities of the SC and the fastigial nucleus (FN) of the cerebellum are identified as the saccade initiator and terminator, respectively. Description of the synaptic properties of the major neural sites involved in the execution of a saccade provides the basis for developing quantitative computational models of the neural network.

2.2.1   SUPERIOR COLLICULUS

The SC initiates the saccade and is considered to translate visual stimuli to motor commands. It includes two important functional regions: the superficial layer and the deep layers (Enderle and Zhou, 2010). The superficial layer is conventionally considered the visual layer that receives the information from the retina and the visual cortex. The deep layers, however, are involved with generation of the desired efferent commands for initiating saccades. It should be noted that the deep layers cause a high-frequency firing that starts 18–20 ms before a saccade, and ends almost when the saccade is complete.

2.2.2   PREMOTOR NEURONS IN THE PPRF

The paramedian pontine reticular formation (PPRF) encompasses neurons that show dominantly increasing burst frequencies of up to 1,000 Hz during the saccade and remain inactive during the periods of fixation. The LLBN and the medium lead burst neuron (MLBN) are the two types of

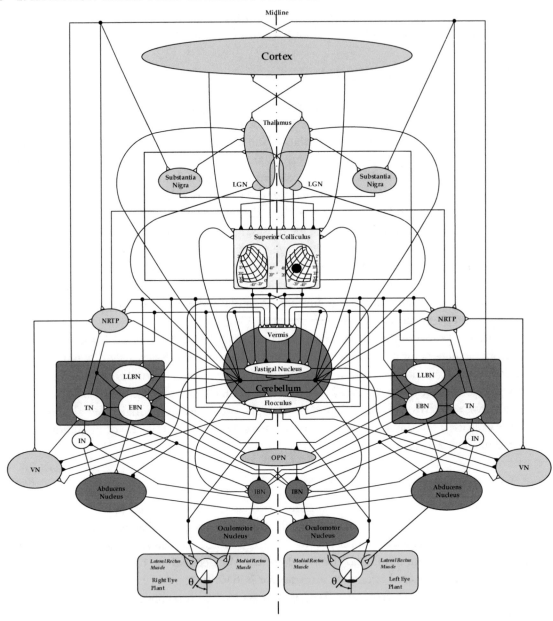

Figure 2.1: The parallel-distributed neural network for generation of a 20° conjugate goal-directed horizontal saccade in both eyes. Excitatory and inhibitory inputs are shown with white and black triangles at the postsynaptic neurons, respectively. This network is an updated network of that proposed by Enderle and Zhou (2010) such that IN mediates between TN and Abducens Nucleus (AN). In addition, the IN is inhibited by the IBN on each side.

burst neurons in the PPRF. The LLBN forms an excitatory synapse to the IBN and an inhibitory synapse to the OPN. Although the biophysical properties of the LLBN are not strongly related to the saccadic characteristics observed in the data, its functionality is essential to the control of the saccades.

There are two types of neurons in the MLBN: the EBN and the IBN. The EBN serves as one of the vital excitatory inputs for the saccade controller. The primary inputs to this neuron are the excitatory input of the SC and the inhibitory input from the contralateral IBN and OPN. This neuron forms excitatory synapses to the TN and the AN. The IBN, on the other hand, controls the firing of the EBN as well as the TN, both of which are on the opposite side of the network to the corresponding IBN. It also inhibits the ON and the IN on the same side as itself. This neuron receives excitatory inputs from the FN of the cerebellum on the opposite side and the LLBN on the same side, and an inhibitory input from the OPN.

2.2.3   OMNIPAUSE NEURON

The OPN inhibits the MLBNs during the periods of fixation, and is inhibited by the LLBN during the saccade. It stops firing about 10–12 ms before the saccade starts, and resumes firing approximately 10 ms before the saccade ends. It receives exclusively inhibitory inputs from the LLBN on either side of the network.

2.2.4   TONIC NEURON

The TN is responsible for keeping the rectus eye muscles steady once the saccade completes. This neuron receives excitatory input from the corresponding EBN and inhibitory input from the opposite IBN. During saccades, the tonic neuron remains silent until the saccade ends. At this point, the tonic neuron generates a signal of variable frequency, depending on how far the eye has moved from its initial position. In particular, the tonic neuron functions as an integrator generating an action potential train whose frequency is directly proportional to the integrated EBN signal.

2.2.5   INTERNEURON

Many excitatory and inhibitory INs in the central nervous system stimulate and control motoneurons. The cerebellum aggregates most of these INs whose functionality depends on the anatomical aspects and properties of their membranes. The IN receives the excitatory and inhibitory inputs from the corresponding TN and IBN, respectively. It consecutively provides the step component to the agonist and antagonist neural controllers.

2.2.6   ABDUCENS NUCLEUS

The burst discharge in the motoneurons resembles a delayed EBN burst signal and is responsible for movement of the eyes conjugately. In motoneurons, the end structure of the axon is connected firmly to the muscle membrane. The AN drives the lateral rectus eye muscle, while also firing in synchrony with the ON from the opposite side. It is excited by the EBN during the saccade and by the IN once the saccade is completed. The IBN on the opposite side inhibits this neural population during the periods of fixation.

2.2.7   OCULOMOTOR NUCLEUS

The ON is solely responsible for the stimulation of the medial rectus eye muscle. This nucleus receives excitatory input from the opposite AN, and inhibitory input from the corresponding IBN.

2.2.8   CEREBELLUM

The cerebellum functions as a time-optimal gating element by using three active sites, namely, the cerebellar vermis (CV), the fastigial nucleus (FN), and the flocculus during the saccade (Enderle and Zhou, 2010). The CV retains the current position of the eye by registering the information on the proprioceptors in the oculomotor muscles and an internal eye position reference. The CV also keeps track of the dynamic motor error used to control the saccade amplitude in connection with the nucleus reticularis tegmenti pontis (NRTP) and the SC. The FN is stimulated by the SC and projects ipsilaterally and contralaterally to the LLBN, IBN, and the EBN on the opposite side of the network. The contralateral FN starts bursting 20 ms before the saccade, while the ipsilateral FN undergoes a pause in firing and discharges with a burst slightly before the saccade completion. The third site, the flocculus, increases the time constant of the neural integrator for saccades with starting locations dissimilar to the primary position. By virtue of the physiological evidence, the cerebellum is responsible for terminating a saccade precisely, with respect to the primary position of the eye in the orbit (Enderle and Zhou, 2010).

2.3   FIRING CHARACTERISTICS OF EACH TYPE OF NEURON

The saccade generator investigated herein is built upon the extant research (Enderle, 1994, Enderle and Engelken, 1995, Enderle, 2002, Enderle and Zhou, 2010, Enderle and Bronzino, 2011, Zhou et al., 2009). The model is 1st-order time-optimal; that is, it does not depend on the firing rate of the neurons to determine the saccade magnitude. Here, we demonstrate features of the dynamics of the saccade neural network to highlight the underlying neurological control processes.

2.3.1   NEURAL ACTIVITY

The structure of the saccade neural network leverages a neural coding so that burst duration is transformed into saccade amplitude under the time-optimal condition. Such coding manifests activities—including the onset of burst firing before saccade, peak firing rate, and end of firing with respect to the saccade termination—for each neuron on the basis of the physiological

Table 2.1: Firing activity features of the important neural sites during an ipsilateral saccade

Neural Site	Burst Onset Before Saccade (ms)	Peak Firing Rate (Hz)	Burst End with Respect to Saccade End
Contralateral SC	20–25	800–1000	Almost the same
Ipsilateral LLBN	20	800–1000	Almost the same
OPN	6–10	150–200 (before and after)	Almost the same
Ipsilateral EBN	6–8	600–1000	~ 10 ms before
Ipsilateral IBN	6–8	600–800	~ 10 ms before
Ipsilateral TN/IN	5	Tonic firing (before and after)	Resumes tonic firing when saccade ends
Ipsilateral AN	5	400–800	~ 5 ms before
Ipsilateral FN	20	Pause during saccade and a burst of 200 Hz near the end of the saccade	Pause ends with burst ~ 10 ms before saccade ends; resumes tonic firing ~ 10 ms after saccade ends
Contralateral FN	20	200	Pulse ends with pause ~ 10 ms before saccade ends; resumes tonic firing ~ 10 ms after saccade ends
Ipsilateral CV	20–25	600–800	~ 25 ms before
Ipsilateral NRTP	20–25	800–1000	Almost the same
Ipsilateral Substantia Nigra	40	40–100	Resumes firing ~40–150 ms after saccade ends

evidence. These characteristics are provided for the neural sites as a framework for our simulations (Enderle and Zhou, 2010). Table 2.1 summarizes the activities in initiation, control, and termination of the burst firing through the neural network, generating a saccade in the right eye.

2.3.2   BURST DISCHARGE MECHANISM

As motoneurons receive excitatory input from the ipsilateral EBN, the burst discharge in them during a saccade is adequately similar to that of ipsilateral EBN. Such burst discharge in the motoneurons is responsible for the movement of the rectus muscles during a saccade. The firing rate trajectory of the EBN is of prime importance in control of such a saccade. The presented EBN model showed a constant plateau of bursting during the second portion of the burst before the decay occurs Enderle and Zhou (2010). Figure 2.2 shows the EBN bursting rate as a fit to the data in Robinson (1981), with (A), and the data in Gancarz and Grossberg (1998), with (B), for three saccades. Note that the interval $[0, T_1]$ depicts the smallest possible interval required for EBN burst according to physiological evidence. The interval $[T_1, T_2]$ represents the duration of the second portion of the burst, by the end of which the EBN drives the motoneurons to move each eye to its destination. The gradual decay in firing occurs in the interval from T_2 until the EBN stops firing. This interval indicates the time it takes for the OPN to resume its inhibition of the EBN. The mechanism for introducing this EBN decay in firing into the axon model is to reduce the firing rate linearly by modifying the channel equations, as described later. Note that the only difference between the three saccades is the duration $T_2 - T_1$. In other words, the saccade magnitude is governed only by the duration of time that the ipsilateral EBN bursts in the interval $[0, T_2]$. We model the EBN firing rate by applying the firing rate trajectory shown in Fig. 2.2B, where a slow linear reduction in firing rate is assumed in the interval $[T_1, T_2]$. This slow drop in firing rate has been reported to be attributed to the IBN inhibition of the LLBN (Gancarz and Grossberg, 1998). We also consider this trajectory for the contralateral SC and FN current stimulation of the ipsilateral LLBN, as shown in Fig. 2.3. These trajectories accord with those of the different simulations in examining the effects of several depolarizing stimulus currents in the EBN axon (Enderle and Zhou, 2010). It should be emphasized at this point how the SC contributes to the optimal control of the saccades by driving the LLBN. The neural activity in the SC is arranged into movement fields that are related to the direction and saccade amplitude (Zhou et al., 2009). The movement fields within the SC are indicators of the number of neurons firing for different small and large saccades (see locus of points on a detailed view of the SC retinotopic mapping in Fig. 2.14 of Enderle and Zhou (2010)).

Neurons active in the SC in Fig. 2.1 are shown with the dark circle, representing the locus of points for a desired 20° saccade. Enderle and Zhou (2010) report that active neurons in the deep layers of the SC generate a sporadic high-frequency burst of activity that varies with time, initiating 18–20 ms before a saccade and ending sometime toward the end of the saccade. However, the exact timing for the end of the SC firing happens quite randomly and can be either before or after the saccade ends. It is implied that the number of cells firing in the LLBN is de-

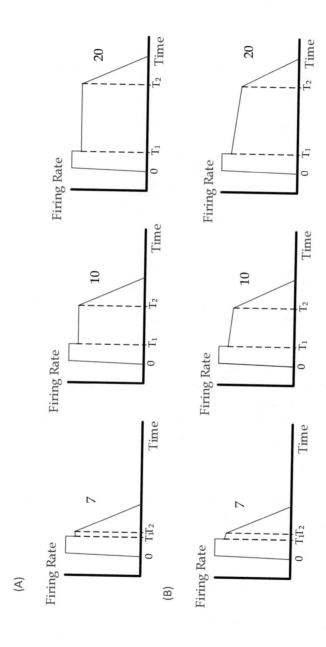

Figure 2.2: Block sketch of EBN firing rates for 7°, 10°, and 20° saccades drawn to match the data. (A) data from Robinson (1981). (B) data from Gancarz and Grossberg (1998).

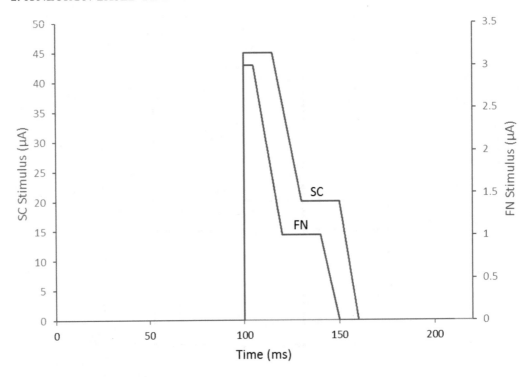

Figure 2.3: The current stimulation trajectories of the ipsilateral LLBN. The current amplitude for the contralateral SC and contralateral FN is chosen based on the burst properties for these two neural sites provided in Table 2.1. As for the contralateral FN stimulus current, a step current of 3 μA is applied at 100 ms, shortly after which a linear decrease in stimulus to 1 μA appears. Subsequently, another step stimulation continues until 140 ms, when a linear reduction occurs until the current is removed at 150 ms.

termined by the number of cells firing in the SC, as long as there is a feedback error maintained by the cerebellar vermis. The number of the OPN cells firing after inhibition from the LLBN determines, in turn, how many EBN cells are released from inhibition. In consequence, the number of EBN cells firing determines the number of motoneurons driving the eyes to their destination.

2.3.3   SEQUENCE OF NEURAL FIRING

The saccade completion involves the evolution of some events in an orderly sequence in the neural sites. Such neural sites are described in Fig. 2.4 via a functional block diagram (Enderle and Zhou, 2010). The output of each block indicates the firing pattern at each neural site manifested during the saccade: the saccade starts at time zero, and T represents the saccade termination.

The negative time for each neural site refers to the onset of the burst before the saccade (see Table 2.1). The neural activity within each block is represented as pulses and/or steps, consistent with the described burst discharge mechanism, to reflect the neural operation as timing gates. Finally, motoneurons innervate rectus muscles in both eyes at the end interaction level of the block diagram.

The following description outlines eight steps required to implement the saccade control strategy in the context of Fig. 2.4. It represents the sequence of events accounted for in Enderle and Zhou (2010), with modifications made in steps 4–7 to indicate the function of local neural integrators (TN and IN) in providing the step of innervation to the moto-neurons:

1. The deep layers of the SC initiate a saccade based on the distance between the current position of the eye and the desired target.

2. The ipsilateral LLBN and EBN are stimulated by the contralateral SC burst cells. The LLBN then inhibits the tonic firing of the OPN. The contralateral FN also stimulates the ipsilateral LLBN and EBN.

3. When the OPN ceases firing, the MLBN (EBN and IBN) is released from inhibition.

4. The ipsilateral IBN is stimulated by the ipsilateral LLBN and the contralateral FN of the cerebellum. When released from inhibition, the ipsilateral EBN responds with a post-inhibitory rebound burst for a brief period of time. The EBN, when stimulated by the contralateral FN (and perhaps the SC), enables a special membrane property that causes a high-frequency burst that decays slowly until inhibited by the contralateral IBN. The burst-firing activity of EBN is integrated through the connection with the TN. The IN follows closely the same integration mechanism as that of the TN.

5. The burst firing in the ipsilateral IBN inhibits the contralateral EBN, IN, and AN, as well as the ipsilateral ON.

6. The burst firing in the ipsilateral EBN causes the burst in the ipsilateral AN, which then stimulates the ipsilateral lateral rectus muscle and the contralateral ON. With the stimulation of the lateral rectus muscle by the ipsilateral AN, and the inhibition of the ipsilateral medial rectus muscle via the ON, a saccade occurs in the right eye. Simultaneously, the contralateral medial rectus muscle is stimulated by the contralateral ON, and, with the inhibition of the contralateral lateral rectus muscle via the AN, a saccade occurs in the left eye. Hence, the eyes move conjugately under the control of a single drive center. During the fixation periods, the INs provide the steady-state tensions required to keep the eyes at the desired destination.

7. At the termination time, the cerebellar vermis, operating through the Purkinje cells, inhibits the contralateral FN and stimulates the ipsilateral FN. Some of the stimulation of the ipsilateral LLBN and IBN is lost because of the inhibition of the contralateral FN. The

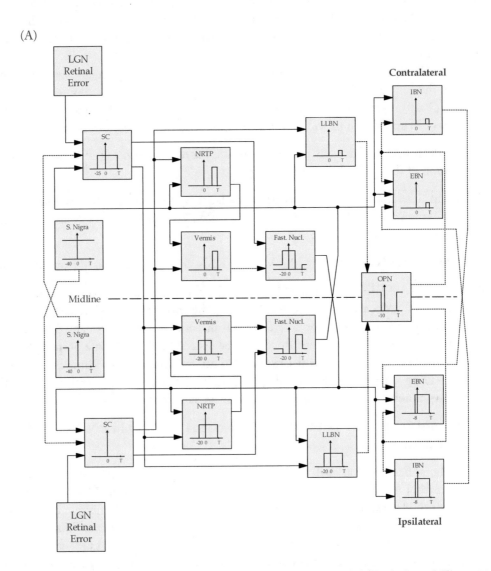

Figure 2.4: A functional block diagram of the saccade generator model (Enderle and Zhou, 2010). Solid lines are excitatory and dashed lines are inhibitory. Each block represents the neural activity of the corresponding neural site as indicated in Table 2.1. (A) Neural pathways from the formation of the lateral geniculate nucleus (LGN) retinal error to the MLBN.

(B)

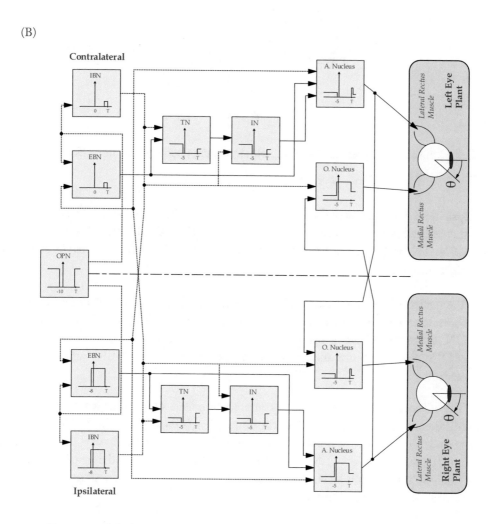

Figure 2.4: *(Continued.)* A functional block diagram of the saccade generator model (Enderle and Zhou, 2010). Solid lines are excitatory and dashed lines are inhibitory. Each block represents the neural activity of the corresponding neural site as indicated in Table 2.1. (B) Neural pathways from the MLBN to the rectus muscles in both eyes.

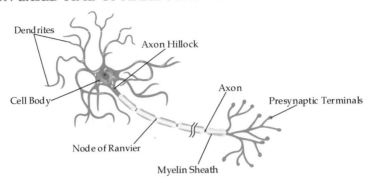

Figure 2.5: A schematic presentation of the different components of a neuron (Enderle and Bronzino, 2011).

ipsilateral FN stimulates the contralateral LLBN, EBN, and IBN. The contralateral EBN then stimulates the contralateral AN. The contralateral IBN then inhibits the ipsilateral EBN, TN, and AN, and the contralateral ON. This inhibition removes the stimulus to the agonist muscle.

8. The ipsilateral FN stimulation of the contralateral EBN allows for modest bursting in the contralateral EBN. This activity then stimulates the contralateral AN and the ipsilateral ON. Once the SC ceases firing, the stimulus to the LLBN stops, allowing the resumption of OPN firing that inhibits the ipsilateral and contralateral MLBN, hence terminating the saccade.

The advances in computational neural modeling have supplied us with abundant information at different structural scales, such as the biophysical (Ghosh-Dastidar and Adeli, 2007, 2009, Mohemmed et al., 2012), the circuit (Enderle and Zhou, 2010, Rossell'o et al., 2009), and the systems levels (Ramanathan et al., 2012). The following includes our modeling of the premotor and motor neurons at the circuit level. We introduce a neural circuit model that can be parameterized to match the described firing characteristics of each type of neuron.

2.4   NEURAL MODELING

A typical neuron embodies four major components: cell body, dendrites, axon, and presynaptic terminals, as shown in Fig. 2.5. The neural cell body encompasses the nucleus, as is true of other cells. Dendrites act as the synaptic inputs for the preceding excitatory and inhibitory neurons. Upon this stimulation of the neuron at its dendrites, the permeability of the cell's plasma membrane to sodium intensifies, and an action potential moves from the dendrite to the axon (Enderle and Bronzino, 2011). The transmission of an action potential along the axon is facilitated by means of nodes of Ranvier in the myelin sheath. At the end of each axon there are presynaptic terminals, from which the neurotransmitters diffuse across the synaptic cleft.

A complete understanding of the properties of a membrane by means of standard biophysics, biochemistry, and electronic models of the neuron will lead to a better analysis of membrane potential response. A neuron circuit model is desired to quantify the saccade-related neural activity, thus reflecting the physiology linked to the dendrite, cell body, axon, and presynaptic terminal of each neuron. Such a model is sketched in this section, together with the description of the modifications on it that are required to populate a neural network for the control of saccades. The saccade neural network includes eight neuron populations at premotor and motor levels as seen in Fig. 2.1:

1. Long lead burst neuron (LLBN),

2. Omnipause neuron (OPN),

3. Excitatory burst neuron (EBN),

4. Inhibitory burst neuron (IBN),

5. Tonic neuron (TN),

6. Interneuron (IN),

7. Abducens nucleus (AN),

8. Oculomotor nucleus (ON).

The saccade circuitry underlies the dynamics of the above eight distinct neurons, each of which contributes to the control mechanism of the saccade. Except for the OPN, the proposed parallel-distributed neural network accommodates two of each of the other neurons in the network. The dendrite model delineated below is adjustable to the stimulation mechanism of all eight neurons. The axon model for all spiking neurons, except the EBN and OPN, adheres to the Hodgkin—Huxley (HH) model. The EBN and OPN are neurons that fire automatically when released from inhibition—these neurons are modeled using a modified HH model (Enderle and Zhou, 2010). The TN integrates its input and is modeled with a FitzHugh—Nagumo (FHN) model under the tonic bursting mode (Faghih et al., 2012). The presynaptic terminal elicits a pulse train stimulus whose amplitude depends on the membrane characteristics of the postsynaptic neuron.

2.4.1 DENDRITE MODEL

The dendrite is partitioned into a number of membrane compartments, each of which has a predetermined length and diameter. Each compartment in the dendrite has three passive electrical characteristics: electromotive force (emf), resistance, and capacitance, as shown in Fig. 2.6. Axial resistance is used to connect the dendrite to the axon.

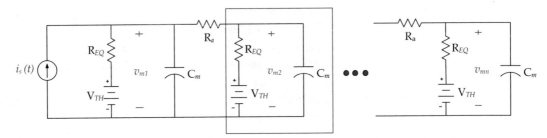

Figure 2.6: The dendrite circuit model with n passive compartments: $i_s(t)$ models the stimulus current from the adjacent neurons to the dendrite. Each compartment has membrane electromotive, resistive, and capacitive properties—V_{TH}, R_{EQ}, and C_m in the second compartment are noted. The batteries in the circuit, V_{TH}, are the Thévenin equivalent potential of all the ion channels. The axial resistance R_a connects each compartment to the adjacent ones (remains unchanged among the neurons). Appropriate values for the membrane resistance and capacitance of the dendrite model are found to match physiological evidence for each neuron.

The presynaptic input to the dendrite is modeled as a pulse train current source (i_s). The node equation for the first dendrite compartment is

$$C_m \frac{dv_{m1}}{dt} + \frac{v_{m1} - V_{TH}}{R_{EQ}} + \frac{v_{m1} - v_{m2}}{R_a} = i_s, \tag{2.1}$$

where v_{m1} is the membrane potential of the first compartment and v_{m2} is the membrane potential of the second compartment. The membrane resistance R_{EQ}, capacitance C_m, and emf V_{TH} characterize each compartment. R_a is the axial resistance.

For all intermediate dendrite compartments there are two inputs: the input from the previous compartment's membrane potential and the input from the next compartment's membrane potential. The node equation for the second compartment is

$$C_m \frac{dv_{m2}}{dt} + \frac{v_{m2} - V_{TH}}{R_{EQ}} + \frac{v_{m2} - v_{m1}}{R_a} + \frac{v_{m2} - v_{m3}}{R_a} = 0, \tag{2.2}$$

where v_{m3} is the membrane potential of the third compartment.

The last dendrite compartment receives just one input from its preceding compartment. The corresponding node equation is

$$C_m \frac{dv_{mn}}{dt} + \frac{v_{mn} - V_{TH}}{R_{EQ}} + \frac{v_{mn} - v_{m(n-1)}}{R_a} = 0, \tag{2.3}$$

where the membrane potential v_{mn} is related to the preceding compartment's membrane potential ($v_{m(n-1)}$) through the axial resistance R_a.

The dendrite model of each neuron is determined by fine-tuning the parametric capacitance and resistance properties of the above defined dendrite model. This parametric adaptation allows for the accommodation of the synaptic transmission in the neural network, as required to stimulate each postsynaptic neuron. Each neuron's dendrite rise-time constant determines the delay to emulate the postsynaptic potential propagation along the dendrite, consistent with the initiation of firing with respect to the saccade onset provided in Table 2.1. For instance, from the EBN's dendrite Thévenin equivalent circuit, nearly five time constants provides the necessary time delay between the OPN's end of firing and the EBN's onset of firing. Table 2.2 includes the membrane resistance and capacitance of the dendrite compartments for each neuron.

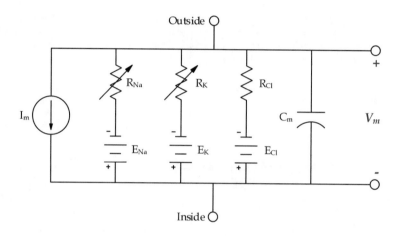

Figure 2.7: The circuit model of an unmyelinated portion of squid giant axon (Enderle and Zhou, 2010). The variable active-gate resistances for Na^+ and K^+ are given by $R_K = 1/\bar{g}_K N^4$ and $R_{Na} = 1/\bar{g}_{Na} M^3 H$, respectively. The passive gates are modeled by a leakage channel with resistance, $R_l = 3.33$ kΩ. The battery is the Nernst potential for each ion: $E_l = 49.4$ V, $E_{Na} = 55$ V, and $E_K = 72$ V.

The initial condition (state) of the capacitor is set to V_{TH}. Computational efficiency accrues when the minimum number of compartments in the dendrite model is required. We chose to include 14 compartments in the dendrite to achieve the desired membrane properties in each type of neuron. For example, the EBN dendritic membrane potential across the first, second, third, and last compartments is illustrated in Fig. 2.8. The farther the compartment is along the dendrite, the smoother its potential response to the pulse train current source. The last compartment of the post-synaptic dendrite (cell body) leads the signal flow to the axon—the site of action potential generation.

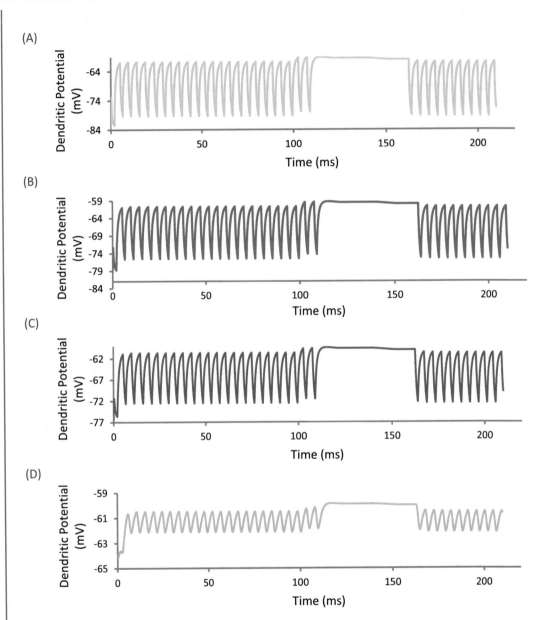

Figure 2.8: The EBN dendritic membrane potential across the different compartments. (A) first compartment, (B) second compartment, (C) third compartment, and (D) last compartment. The membrane parameter values are: $V_{TH} = -60$ mV, $C_m = 0.45$ μF, $R_{EQ} = 3.1$ kΩ, and $R_a = 100$ Ω.

2.4.2   AXON MODEL

Quite a few circuit models can be considered to reproduce the electrical properties of an axon in simulation of SNNs. The choice has to set forth a compromise between several factors, such as physiological realism, computational cost, complexity, accuracy, and scalability. The Hodgkin-Huxley (HH) model of the axon serves as the basis for the neurons modeled here—only the EBN and the OPN are based on a modified HH model. As elaborated later, this modification leads these neurons to fire automatically at high rates after releasing from inhibition, given minor stimulation. The HH model is developed to describe the membrane potential at the axon hillock caused by conductance changes (Enderle and Zhou, 2010). The circuit diagram of an unmyelinated portion of squid giant axon is illustrated in Fig. 2.7. The node equation that expresses the membrane potential V_m as a function of stimulus current I_m from the dendrite and voltage-dependent conductances of the sodium and potassium channels is

$$\bar{g}_K N^4 (V_m - E_K) + \bar{g}_{Na} M^3 H (V_m - E_{Na} + \frac{(V_m - E_l)}{R_l} + C_m \frac{dV_m}{dt} = I_m, \qquad (2.4)$$

where

$$\frac{dN}{dt} = \alpha_N (1 - N) - \beta_N N,$$

$$\frac{dM}{dt} = \alpha_M (1 - M) - \beta_M M,$$

$$\frac{dH}{dt} = \alpha_H (1 - H) - \beta_H H,$$

$$\bar{g}_K = 36 \times 10^{-3} S, \qquad \bar{g}_{Na} = 120 \times 10^{-3} S.$$

The coefficients in the above 1st-order system of differential equations are related exponentially to the membrane potential V_m, i.e.,

$$\alpha_N = 0.01 \times \frac{V + 10}{e^{\left(\frac{V+10}{10}\right)} - 1} \quad ms^{-1}, \qquad \beta_N = 0.125 e^{\left(\frac{V}{80}\right)} \quad ms^{-1},$$

$$\alpha_M = 0.1 \times \frac{V + 25}{e^{\left(\frac{V+25}{10}\right)} - 1} \quad ms^{-1}, \qquad \beta_M = 4e^{\left(\frac{V}{18}\right)} \quad ms^{-1}, \qquad (2.5)$$

$$\alpha_H = 0.07 e^{\left(\frac{V}{20}\right)} \quad ms^{-1}, \qquad \beta_H = \frac{1}{e^{\left(\frac{V+30}{10}\right)} + 1} \quad ms^{-1},$$

$$V = V_{rp} - V_m \quad mV$$

where the resting potential V_{rp} is −60 mV.

The neural firing rate of all the bursting neurons has been adjusted to meet the peak firing rate requirement in Table 2.1. This adjustment intends for each neuron to contribute to the generation of the saccade by mimicking the required physiological properties (Enderle and Zhou, 2010). To this end, the right-hand side of the N, M, and H differential expressions in Eq. (2.4) is multiplied by appropriate coefficients to achieve the desired peak firing rates. For instance, the required coefficient for the EBN is 35,000; thereby it presents a peak firing rate at 1,000 Hz.

It should be pointed out that the above equations of the basic HH model of the axon have been used for all the bursting neurons, except for the EBN and the OPN. For these latter neurons, the modified HH model is used to change the threshold voltage from –45 mV to –60 mV. Enderle and Zhou (2010) illustrated experiments in which this variation caused EBN to fire autonomously without the existence of any excitatory stimulus. From their description of the dominant effect of the sodium channel current on the changes in the threshold voltage at the beginning of the action potential, the threshold voltage in the EBN axon model is changed by modifying the α_M equation to

$$\alpha_M = 0.1 \times \frac{V + 10}{e^{\left(\frac{V+10}{10}\right)} - 1} \quad \text{ms}^{-1}. \tag{2.6}$$

The OPN axonal threshold voltage of firing has been adjusted following the same modification by the above equation. This alteration of the threshold voltage for the EBN and the OPN enables them to fire spontaneously without any significant depolarization from peripheral current stimuli. Table 2.2 lists the firing threshold voltage and the coefficient required to adjust the peak firing rate for each bursting neuron.

The axon transfers an action potential from the spike generator locus to the output end, the synapse. The transmission along the axon thus amounts to introducing a time delay, after which the action potential appears at the synapse.

2.4.3 SYNAPSE MODEL

When the action potential appears at the synapse, packets of neurotransmitter are released. This is modeled by excitatory or inhibitory pulse train stimuli to stimulate the dendrite of the postsynaptic neuron more realistically. Current-based synapse models offer significant analytical convenience when describing how a postsynaptic current pulse is triggered by an action potential in very large SNNs (Wong et al., 2012). As these models disregard the voltage-dependent property of the postsynaptic currents, for the networks with both the interspike intervals and the burst onsets of the neurons uniformly distributed, they are preferred to the conductance-based synapse models. Following the concepts of the current-based synapse models, the amplitude and width of synaptic pulses are tweaked in simulation runs to provide the desired postsynaptic behavior in the interconnected neurons. The width is constrained by the two points at which the action potential crosses a constant level of the axonal potential. In that sense, the synapse can be thought of as a voltage-to-frequency converter that releases a pulse train output. Figure 2.9 shows a number of action potentials and the synaptic current pulses of the EBN toward the end of the burst-firing interval. Note that the time delay between each action potential and the corresponding current pulse is evident.

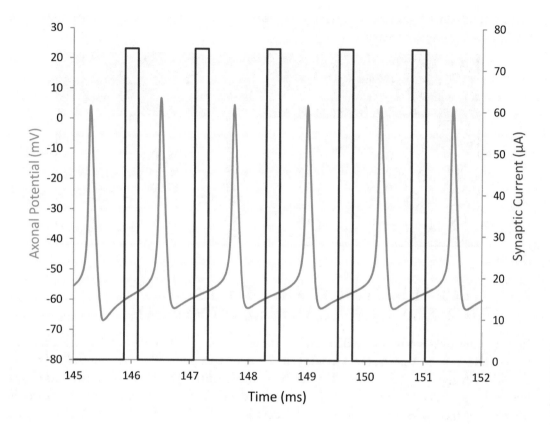

Figure 2.9: A train of action potentials and current pulses reflecting the synaptic mechanism of the EBN. Each current pulse shows a time delay with respect to the corresponding action potential, due to the transmission delay along the axon.

In addition to the transmission time delay along the axon, all chemical synapses introduce a small delay before the generation of postsynaptic potentials from an input excitatory or inhibitory pulse train. This delay accounts for the time required for the release of neurotransmitters and the time it takes for them to distribute through the synaptic cleft. This small synaptic delay was taken into effect by increasing the rise time constant of the subsequent postsynaptic dendritic compartments.

As indicated, the amplitude and width of synaptic current pulses for each neuron are uniquely chosen in order that the postsynaptic neurons exhibit the desired behavior. Table 2.2 includes such amplitude of the synaptic current pulses. This table presents the differences (dendritic, axonal, and synaptic) among eight distinct neurons whose realization is important in the neural circuitry for time-optimal control of the saccade.

Table 2.2: Parametric realization of eight distinct neurons in terms of dendritic, axonal, and synaptic behaviors in the proposed neural circuitry

| Neuron | Dendrite | | Axon | | Synapse |
	Capacitor (μF)	Resistor (kΩ)	Firing threshold (mV)	Coefficient	Pulse amplitude (μA)
LLBN	0.5	3.75	−45	18,000	20
OPN	1.0	6.3	−60	1,800	45
EBN	0.45	3.1	−60	35,000	75
IBN	0.35	4.5	−45	15,000	65
AN	0.35	5.5	−45	17,000	55
ON	0.45	4.0	−45	17,000	55
TN	0.35	4.5	NA	NA	10
IN	0.4	4.5	NA	NA	10

2.5 NEURAL STIMULATION OF THE LINEAR HOMEOMORPHIC MODEL OF MUSCLE

A neural controller is often defined to describe the relationship between the neuron firing rates and the eye orientation. A horizontal saccade can be driven by a pulse-slide-step controller (Zhou et al., 2009). A rigorous mathematical representation of realistic physiological constraints showed that such controller has the capability to preserve the desired neural activities during saccades. The primary focus of our development of the saccade circuitry is to provide motoneuronal control signals, based on the sequence of burst firing indicated in Subsection 2.3.3, to the time-optimal controller. Figure 2.10 illustrates the block diagram of the neural system that includes important neurons at the premotor and motor stages of the neural network, the time-optimal controller, and the oculomotor plant. The intent of the system is the time-optimal controller that has been proven to be adaptable to the physiological and anatomical limitations (Bahill et al., 1980, Enderle and Wolfe, 1988a).

The time-optimal controller model is investigated to obtain the saccadic eye movement model solution that drives the eyeball to its destination for different saccades. As delineated in Section 1.2, the saccadic eye movement model solution is characterized by the realization of the agonist (Eq. (1.29)) and antagonist (Eq. (1.30)) controller models, thereby providing the active-state tensions as inputs to a linear homeomorphic model of the oculomotor plant, as shown in Fig. 2.10.

The agonist controller is a 1st-order pulse-slide-step neuronal controller that describes the agonist active-state tension as the low-pass filtered neural stimulation signal (Eqs. (1.29) and

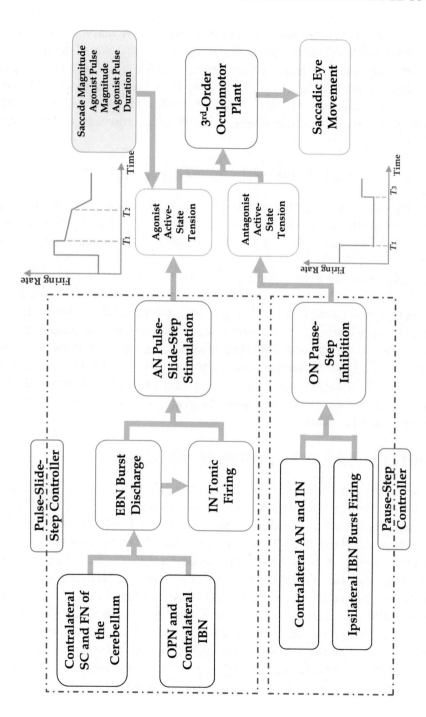

Figure 2.10: A complete block diagram representation of the system of the interaction of motoneuronal stimulation signals and the time-optimal neural controller to drive the oculomotor plant in generation of an ipsilateral saccade.

(1.31)). Under the time-optimal control strategy, the agonist pulse duration determines the saccade magnitude, and the agonist pulse magnitude governs the dynamics of the agonist controller, as depicted in Fig. 2.10. The neural stimulation signal is the firing rate of the ipsilateral AN and that of the contralateral ON. In contrast, the antagonist muscle is unstimulated by a pause during the saccade, and remains fixed by a step input to keep the eyeball at its destination. To serve this purpose, a 1st-order pause-step neuronal controller is defined (Eqs. (1.30) and (1.32)). For the normal saccades with no post-inhibitory rebound burst firing in the antagonist motoneurons, we re-write Eq. (1.32) to

$$\tau_{ant} = \tau_{tde}u(t - T_1) - u(t - T_3) + \tau_{tac}u(t - T_3), \tag{2.7}$$

where the antagonist time constant is described by two step functions, introducing the antagonist deactivation time constant, τ_{tde}, and the activation time constant, τ_{tac}. T_1 takes into account the latent period, and T_3 is the onset of the change to the step component necessary to keep the eyeball steady at its destination.

Ultimately, the inputs to the muscle model are the agonist active-state tension F_{ag} and the antagonist active-state tension F_{ant}, obtained from the above neural controllers (see Eq. (1.26) and Fig. 1.3). The analytical solutions for both F_{ag} and F_{ant} were yielded in Chapter 1, and different saccade characteristics were shown to be very well matched to the experimental data. The neural stimulation analysis in this chapter differs from that of Chapter 1 in that the neural firing rates for the agonist and antagonist motoneurons are not estimated herein, eliminating the need for using any system identification technique. No empirical parameters are involved herein, other than the parameters of the investigated oculomotor plant for humans (Table 1.2). The simulation specifications and results follow.

2.6 NEURAL SYSTEM IMPLEMENTATION

We have investigated three large saccades: 10°, 15°, and 20°. At the neural circuit level, each neuron consists of 14 dendrite compartments with membrane properties included in Table 2.2. The determination of the rise time constant for each neuron's dendrite plays a vital role in the integration of current pulses at the synapse. The onset delay before the saccade, peak firing rate, and burst termination time for the different neuron populations are chosen according to Table 2.1. Analyses of the dendritic membrane potentials were performed with the NI Multisim circuit simulator, and the neural network was simulated in the MATLAB/Simulink 8.0 environment. The modular programming and test of each individual neuron were achieved to constitute our Simulink model of the system of neurons at the highest level of the hierarchy. This implementation, in particular, intends to determine if all the timing requirements are achieved for each module. If a module was yet to satisfy its dynamical features at any stage of the implementation, it was modified and resimulated. More specifications about this implementation follow.

2.6.1   SIMULINK PROGRAMMING

Programming schemes implementing the artificial neural networks (not as the one we described here) mostly use look-up tables to simulate the neurons. The look-up table for each neuron conveys the information about its connections, input weight values, transfer function, and output equation, to describe the entire neurodynamics. A physiologically based model of the neuron is modeled herein, for which a program simulates the underlying membrane differential equations. The main advantage of this neural system is that it offers memory efficiency in allocating the neural activity to each neuron. The information about the neural processing elements (merely the invoked file of parameters listed in Table 2.2) is stored for each neuron. This neural network programming eliminates the barrier of computational cost noticeably. The program is developed in modular structures, thus allowing for analyzing each module to verify whether or not it meets the desired dynamic performance.

The first step to create the block diagram program of any system is to obtain its quantitative mathematical models. For the linear time-invariant dynamic systems, the input-output relationship can be derived in the form of transfer functions. Within the Simulink's block-oriented structure, the transfer function blocks can then be arranged into block diagrams, which are capable of showing the system interconnections graphically. Block diagram representation for a program structure can be implemented in the form of functional modules. As a consequence, each module can be individually developed, tested, and debugged. Finally, when all of the modules meet the desired dynamic performance, they can be linked together to form the main, functioning program of the top-level system. The ode23t solver with variable-step time resolution of the simulator is used to exercise the real-time operation of the neuron model. The program stop time is 210 ms. In the following, we provide an illustration of this top-level system together with its key subsystems, which allows for a review of many concepts presented in this chapter.

The modular structure of the main program of the neural network is presented in Fig. 2.11. This figure demonstrates the program for the block diagram representations of the neural network shown in Fig. 2.4. The program steps through the execution of its modules in an orderly sequence, such that a series of handshake events occurs, as presented in Section 2.3. The SC and FN modules thus have the highest order of execution, the LLBN the next highest, and so on, down to the motoneurons. The time a module takes to execute the synaptic stimuli agrees with the timing properties of the burst firing listed in Table 2.1.

There are a total of 415 blocks (input control ports, processing elements, and output ports) in the main program. Shown in Fig. 2.12 is the block diagram of modules for the EBN. Each module (dendrite, axon, and synapse) depicts a subsystem that is developed separately from the main program. We now engage in a closer look at a number of major properties of the program for each module.

Figure 2.13A shows the program implementing the EBN first dendrite compartment. Recall that Eq. (2.1) is the differential equation that describes the dynamics of the membrane potential of the first compartment. When each synaptic signal flows to the postsynaptic neuron,

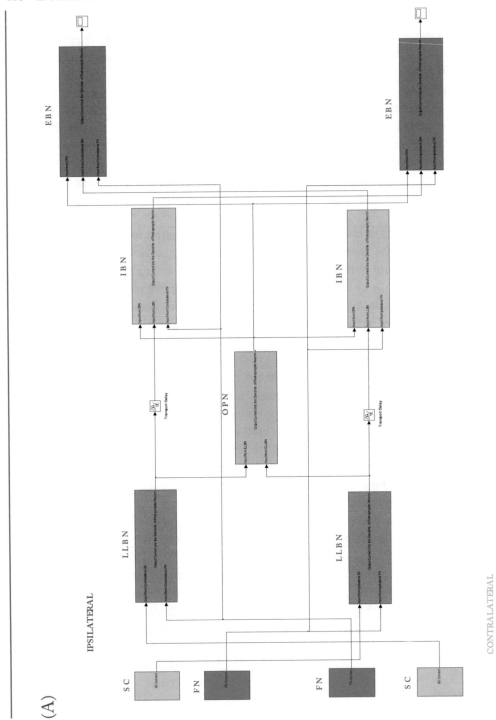

Figure 2.11: The top-level program of the saccade neural network. (A) From the FN and SC modules to the MLBN modules.

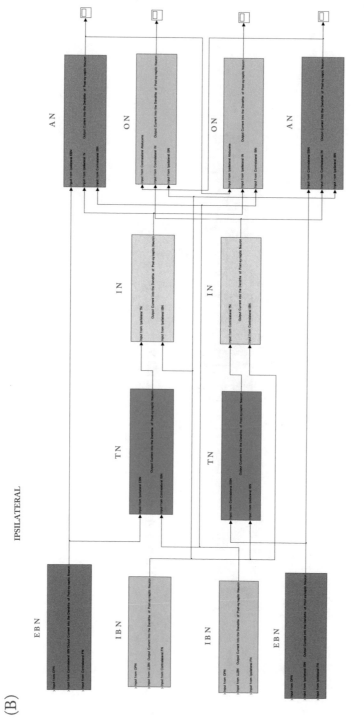

Figure 2.11: *(Continued.)* The top-level program of the saccade neural network. (B) From the MLBN modules to the motoneurons modules.

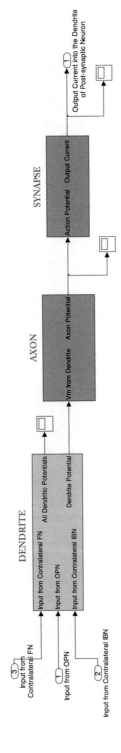

Figure 2.12: Modules of the neural network program for the EBN.

this signal is superimposed with the others to produce a pulse train current source. In the view of this, it is clear that the summation block interfaces the postsynaptic signals with the dendrite compartments. The sign of the input from the synapses to the EBN depends on whether the synapse is excitatory or inhibitory. The output of the integrator is the membrane potential (one of the system's state variables). The block diagram representations of interconnections of the EBN dendrite compartments are depicted in Fig. 2.13B. Evidently, each intermediate compartment receives one input from the previous compartment's membrane potential and the other from the next compartment's membrane potential. Figure 2.13C displays the program for the EBN fourth dendrite compartment. The underlying differential equation is Eq. (2.2) that expresses the dynamics of the membrane potential of each intermediate compartment. The program for the EBN last dendrite compartment, whose descriptive of dynamics is Eq. (2.3), is shown in Fig. 2.13D.

The axon module is the core drive module of the program. The saccade-induced spiking activities at the premotor level are modeled with an HH model for the bursting neurons (Enderle and Zhou, 2010). The tonic spiking behavior of the TN/IN is implemented by a modified FHN model as well (Faghih et al., 2012). The program for the EBN axon module is presented in Fig. 2.14A. The underlying differential equation is Eq. (2.4) that conveys the dynamics of the membrane potential. The program for the M differential equation, with the parameters defined in Eq. (2.5), is depicted in Fig. 2.14B. After the simulation is executed at this level, the results of simulation indicate if the burst activity timing and peak rate are as desired for each neuron. Recall that transmission along the axon introduces a time delay, subsequent to which an action potential appears at the synapse.

At the synapse level, each neuron sends out a pulse train stimuli to the postsynaptic neurons in the context of the neural connections in Fig. 2.11. Synaptic connections between functionally modeled neurons are modeled following a current-based synapse scheme (Wong et al., 2012). The EBN synapse module of the program is displayed in Fig. 2.15. The switch block uses a threshold to convert each action potential to a current pulse with the amplitude defined in Table 2.2. The output pulse width can be altered by changing the threshold.

The program, including the neural controllers and the oculomotor plant, was previously presented in Example 1.1. The programs for the agonist and antagonist neural controllers are presented in Fig. 2.16. From the burst activity of the saccade neural network, the agonist neural input is the firing rate of the ipsilateral AN, and the firing rate of the ipsilateral ON is the antagonist neural input. As presented in Chapter 1, the estimation of these two neural inputs was achieved and programmed in Fig. 1.6. The obtained active-state tensions from Fig. 2.16 drive the program depicted in Fig. 1.6B.

2.6.2 CONTROL SIMULATION RESULTS

Table 2.3 includes the duration of the burst (agonist pulse) for the three different saccades in this work. Notice that the latent period is not zero in our simulations. The saccades start at 120 ms, and they terminate solely after the duration of the burst under the time-optimal control strategy. The

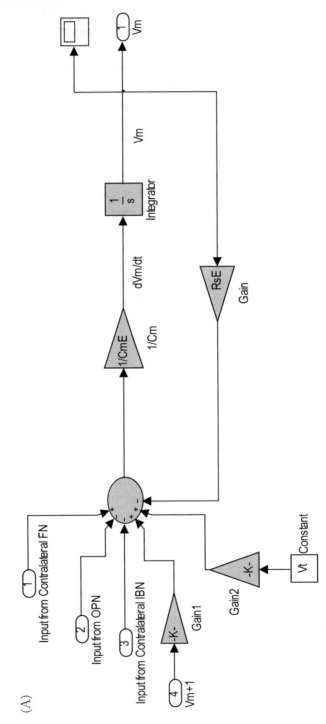

Figure 2.13: The modules of the program for the EBN dendritic compartments. (A) The initial compartment.

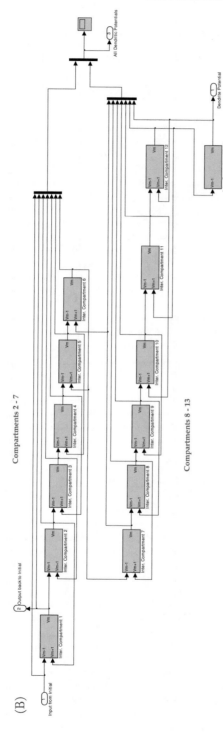

Figure 2.13: *(Continued.)* The modules of the program for the EBN dendritic compartments. (B) Blocks of the intermediate compartments (note the communication layout of the compartments).

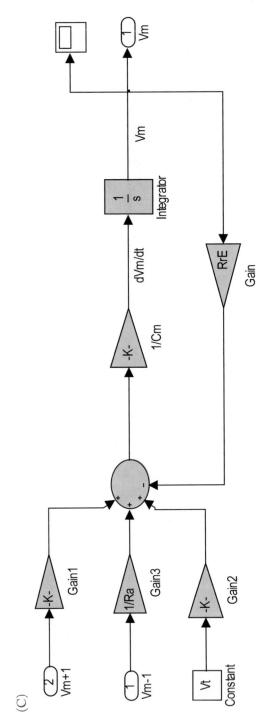

Figure 2.13: *(Continued.)* The modules of the program for the EBN dendritic compartments. (C) The fourth compartment block units.

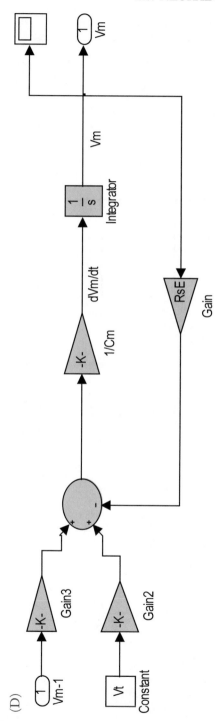

Figure 2.13: *(Continued.)* The modules of the program for the EBN dendritic compartments. (D) The last compartment block units.

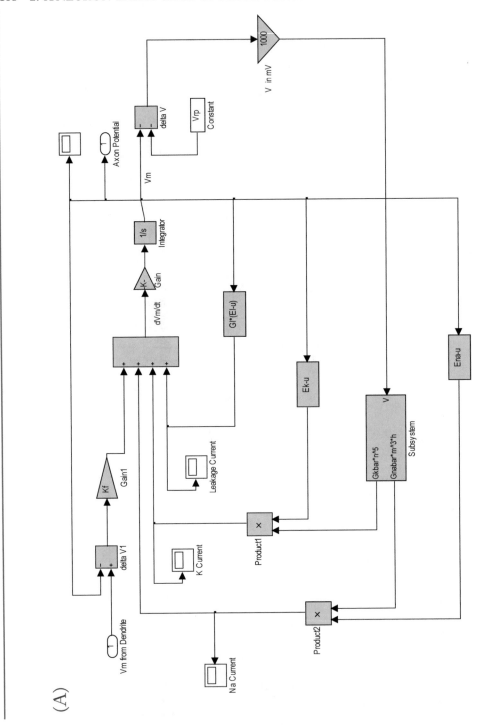

Figure 2.14: The EBN axon module of the program. (A) The main module including the subsystem of N, M, and H coefficients.

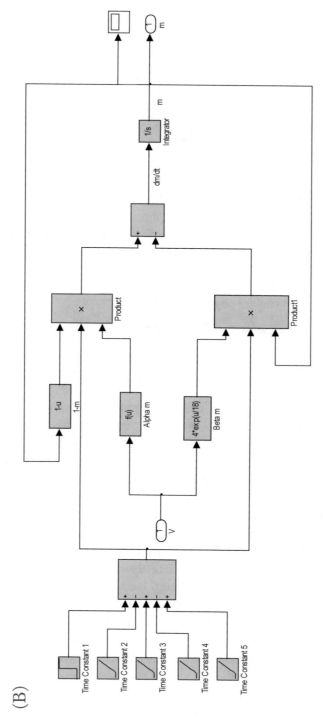

Figure 2.14: (*Continued.*) The EBN axon module of the program. (B) The *M* differential equation of the sodium current block units.

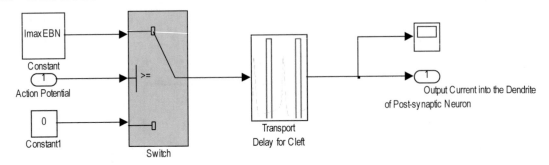

Figure 2.15: The EBN synapse module of the program.

selection of the duration of the burst is in accord with the saccade duration-saccade magnitude characteristic of the main-sequence diagrams (Enderle and Zhou, 2010).

Table 2.3: Time-optimal control of the saccade magnitude with the duration of the burst firing

Saccade Magnitude (Degrees)	Agonist Pulse Duration (ms)
10	50
15	56
20	65

For sample illustrations, the plots of dendritic membrane potential (first column), axonal membrane potential (second column), and synaptic current pulse train (third column) for the burst neurons and the IN in generation of the 10° saccade are shown in Fig. 2.17. Recall that the train of action potentials is converted to a train of the current pulses in the presynaptic terminal of the neuron to provide excitatory or inhibitory input to the succeeding neurons based on the neural connections in Fig. 2.1. This current pulse flows through the postsynaptic dendritic compartments of the latter neurons, thus providing the smooth postsynaptic potentials to prime the axonal compartment. It is evident that, upon the increasing of the stimulus current pulse amplitude, the depolarization of the postsynaptic membrane intensifies.

Duration of burst firing is set to evoke the desired saccades with pulse train synaptic stimuli slightly before the onset of ipsilateral or contralateral saccades (see Table 2.3). It is of interest to note that the burst onset and offset for each premotor neuron in Fig. 2.17 agrees with its place within saccadic circuitry's hierarchical processing order in generating the final motoneuronal signals. From the dendritic potentials in Fig. 2.17, it also appears that the synapse propagation raises different excitatory and inhibitory postsynaptic potentials in the dendritic compartments of each postsynaptic neuron. One can realize that, in view of the trajectory of changes in the membrane potential among the compartments, each postsynaptic neuron, in turn, can either become closer to firing an action potential chain, or inhibited from firing.

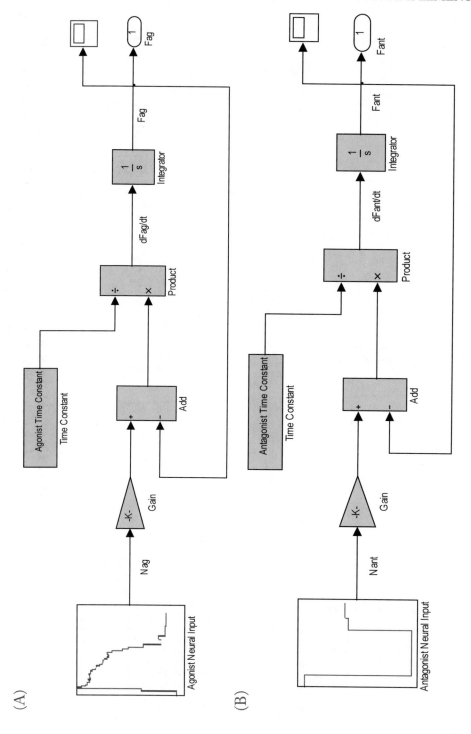

Figure 2.16: The time-optimal neural controller program. (A) The ipsilateral agonist neural controller program. (B) The ipsilateral antagonist neural controller program.

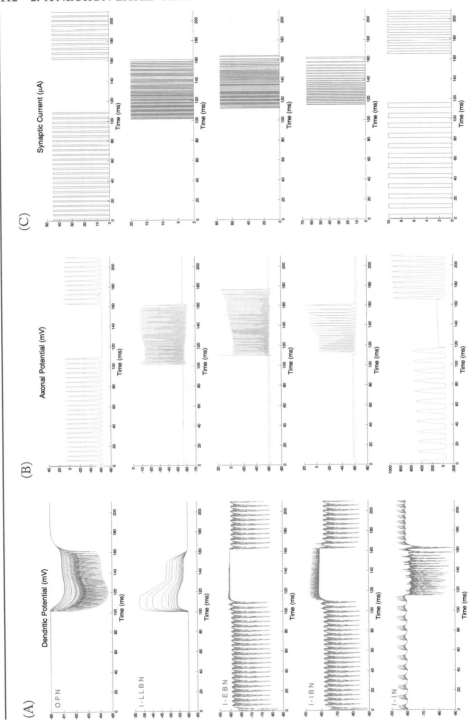

Figure 2.17: The dendritic membrane potential in mV (A), axonal membrane potential in mV (B), and synaptic pulse current train in μA (C) of five neurons in a 10° saccade neural controller. OPN (top), ipsilateral LLBN, EBN, IBN, and IN are shown in order. Each neuron fires in harmony with the others in generating this saccade (saccade onset: 120 ms).

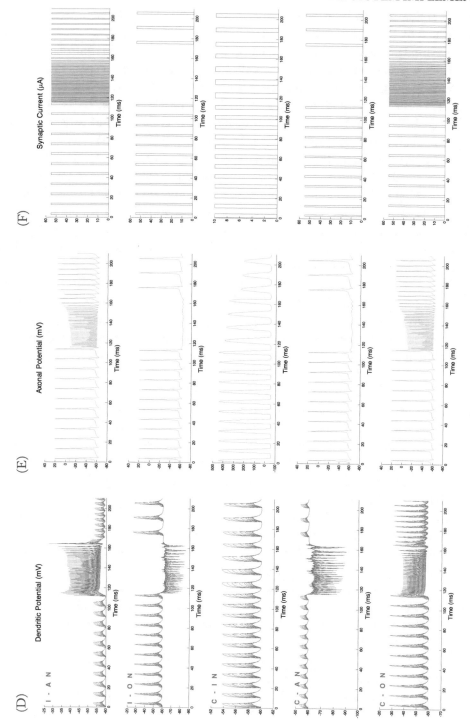

Figure 2.17: *(Continued.)* The dendritic membrane potential in mV (D), axonal membrane potential in mV (E), and synaptic pulse current train in μA (F) of five neurons in a 10° saccade neural controller. Shown in consecutive rows are ipsilateral AN and ON, as well as contralateral IN, AN, and ON.

It is worth noting that the ipsilateral LLBN membrane response is different from the rest, since it is stimulated by the contralateral SC current pulse, as shown in Fig. 2.3. Note that the EBN serves as the fundamental excitatory input for the analysis of the saccade controllers. When the ipsilateral EBN is weakly stimulated by the contralateral FN, it renders a special membrane property that tends to a high-frequency burst mechanism until inhibition from the contralateral IBN and the OPN. The burst-firing trajectory of the ipsilateral EBN for saccades of all sizes is presented in Fig. 2.18. It can be seen that this neuron starts burst firing at very high levels approximately 8 ms before the saccade starts (see Table 2.1). The onset of the second portion of the burst in all cases is 125 ms. The gradual decay in firing occurs in the interval from this instant until approximately 10 ms before the EBN stops firing. The mechanism for modeling this EBN decay in firing in the axon model is to reduce the firing rate linearly by modifying the channel equations, as mentioned previously. These exemplified trajectories are similar to those of the reference illustrated in Fig. 2.2B. It is noteworthy that the only difference between the three saccades is the duration of the second portion of the burst, by the end of which the EBN drives the motoneurons to move each eye to its destination.

The EBN makes excitatory synapses to the TN and AN. It is also noted that the IN firing rate follows that of the TN and is determined based on the current position of the eye before completion of the saccade. In this context, the agonist and antagonist active-state tensions during the periods of fixation are found as functions of eye position at steady-state (see Table 1.2). Obviously, the burst-tonic firing activity of the ipsilateral AN and contralateral ON reflects the burst firing of the ipsilateral EBN and the tonic firing of the IN.

Presented in Fig. 2.19 are the ipsilateral agonist and antagonist burst-tonic firing rates with their respective active-state tensions based on the agonist and antagonist controller models for the three saccades. It is of interest to note that the firing rate of each AN in all scenarios does not vary as a function of saccade magnitude. This provides evidence that the proposed time-optimal controller is quite capable of mimicking the physiological properties of the saccade by merely changing the duration of the agonist burst. The obtained agonist-antagonist firing patterns fairly well match those of the estimated patterns, using the system identification approach given in Section 1.4. In particular, the firing trajectory of the neural input to the agonist controller approximates the burst-tonic data during the pulse and slide phases accurately.

The ipsilateral control simulation results of eye position for the three different saccades under the time-optimal control strategy are demonstrated in Fig. 2.20. The parametric saccadic oculomotor plant for humans has been used (see Table 1.2). From the saccade velocity profile, the peak velocity is found to be $223°s^{-1}$ for the 10° saccade, $277 °s^{-1}$ for the 15° saccade, and $322 °s^{-1}$ for the 20° saccade. It is noteworthy that the investigated oculomotor plant does not considerably influence the main-sequence diagrams, as envisioned by Zhou et al. (2009).

Comparison of the obtained saccade characteristics with the analytical solutions in Section 1.7 demonstrates remarkable consistency. It is noted, however, that, even for the saccades of the same magnitude, there could be recognizable differences in the latent period, time to peak

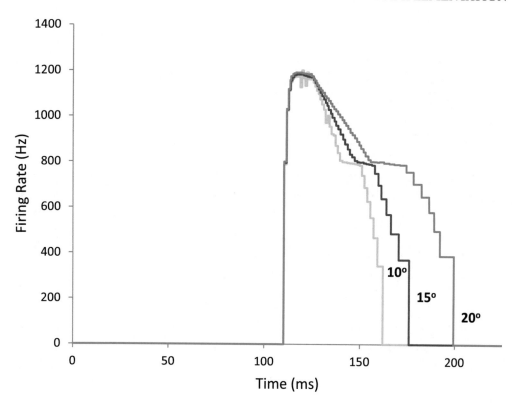

Figure 2.18: Firing trajectories of the ipsilateral EBN for the three saccades. The duration of the second portion of the burst is key in determining the saccade magnitude.

velocity, peak velocity, and peak acceleration. Hence, it is known that saccades of the same magnitude usually exhibit different trajectories. It proves fundamental, nonetheless, that the time-optimal controller fairly well accommodates this variability. The entire eye movement trajectories (position, velocity, and acceleration) on the contralateral side were in close agreement with their corresponding ipsilateral signals for all of the conjugate saccades. From these results, it follows that the agonist burst duration uniquely controls the binocular saccade magnitude under the time-optimal control strategy. The burst duration is found to be correlated to the MLBN duration of burst firing from the extracellular single-unit recordings (Sparks et al., 1976).

As evident by different firing rate trajectories for the EBN, this neuron has characteristics that are tightly coupled to the saccade. For the three saccades examined herein, the initial duration of the EBN firing remained constant among them. However, the duration of the second portion of the burst discharge (gradual drop) varied among them, based on the entire duration of the burst firing in Table 2.3. As indicated in Table 2.1, the EBN firing lags behind the saccade by 6–8 ms,

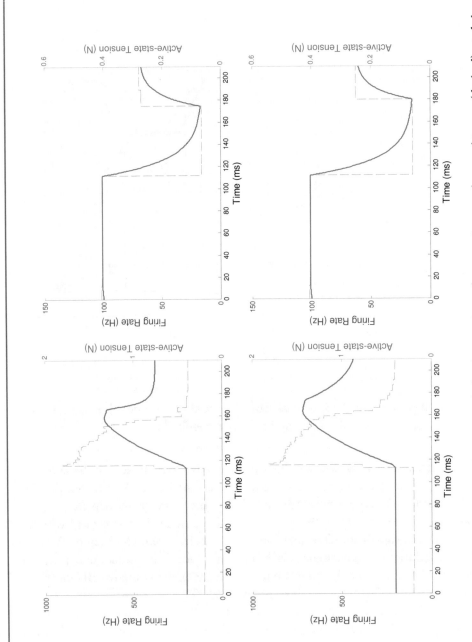

Figure 2.19: The ipsilateral neural stimulation signals for the agonist and antagonist neural control inputs (dashed), and the corresponding active–state tensions (solid) plotted on the same graph. (Top) 10° saccade and (bottom) 15° saccade, The agonist and antagonist controller models provide the active-state tensions to the linear homeomorphic model of the oculomotor plant.

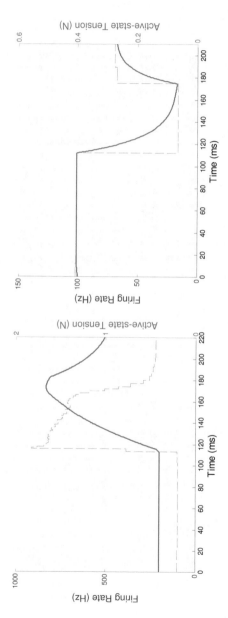

Figure 2.19: *(Continued.)* The ipsilateral neural stimulation signals for the agonist and antagonist neural control inputs (dashed), and the corresponding active-state tensions (solid) plotted on the same graph in a 20° saccade. The agonist and antagonist controller models provide the active-state tensions to the linear homeomorphic model of the oculomotor plant.

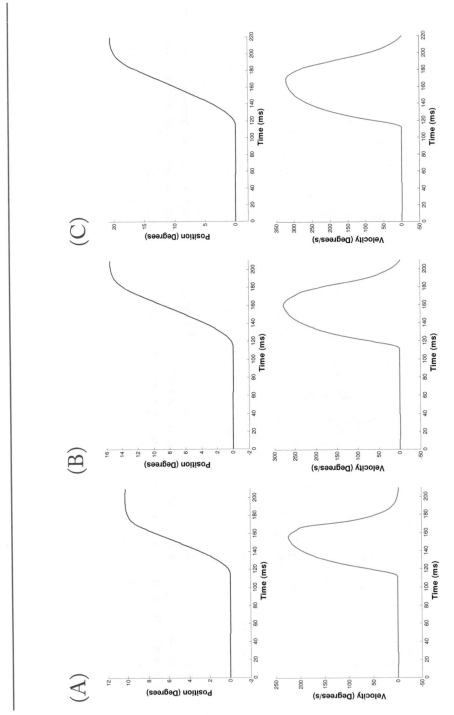

Figure 2.20: The ipsilateral control simulation results (position, velocity, and acceleration) for saccades generated by the proposed 1st-order time-optimal neural saccade controller used in a 3rd-order linear muscle model. (A) 10° saccade, (B) 15° saccade, and (C) 20° saccade. Note that the saccade onset is 120 ms for all cases, but the end of each saccade differs from the others.

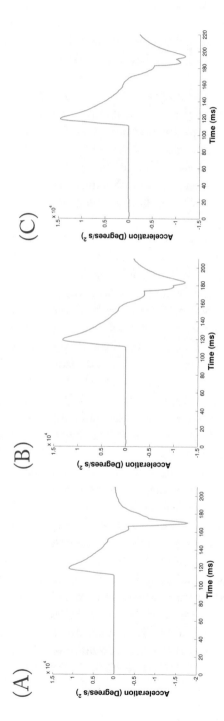

Figure 2.20: *(Continued.)* The ipsilateral control simulation results (position, velocity, and acceleration) for saccades generated by the proposed 1st-order time-optimal neural saccade controller used in a 3rd-order linear muscle model. (A) $10°$ saccade, (B) $15°$ saccade, and (C) $20°$ saccade. Note that the saccade onset is 120 ms for all cases, but the end of each saccade differs from the others.

whereas the AN starts burst firing 5 ms before the saccade. Finding the dendrite parameters for both of these neurons in meeting the required onset time delay was tedious. Moreover, the AN peak firing rate at the beginning of the pulse period showed dependency on the EBN peak firing rate, necessitating the use of corresponding coefficients to change the initial firing rate of the basic HH model (see Table 2.2).

Implementing the OPN dendrite and synapse models in order that this neuron stops inhibiting the EBN about 10 ms before the saccade, and resumes its inhibition almost when the saccade ends, was subject to numerous parameter tunings. Without this coordination in timing of the burst firing in the EBN, this neuron can show the rebound burst-firing activity. This rebound burst, in turn, causes the saccade to deviate from the normal characteristics, as explored later in more detail. It also was vital that the end of the IBN inhibition of the antagonist motoneurons coincides with the resumption of tonic firing in them such that no deviation from the normal saccade is present.

Coubard (2013) reviewed different lines of controversy in research between the proponents of binocular coordination of eyes vs. those of monocular coordination during combined saccade-vergence eye movements. It is suggested that, in order to fully respond to target displacements in all feasible depths and directions, saccade-vergence commands can be simultaneously processed by both eyes, as well as received individually in monocular fashion, especially in view of the neurophysiological manifestations. The treatment to modeling the pure saccades on the basis of the local feedback model (Zee et al., 1992) has been the focus of attention. In this model, a conjugate saccadic velocity command, derived from the saccade burst neuron model, is modulated through local filters to provide the oculomotor motoneurons with the pulse, slide, and step of innervation. In the generation of an ipsilateral saccade, the pulse force of the premotor neurons is attributed to the burst discharge within the PPRF, whereas their step force is related to the burst discharge in the bilateral nucleus prepositus hypoglossi and in the medial vestibular nucleus (Scudder et al., 2002). The OPNs tonically inhibit the premotor neurons as early as the saccade terminates. The premotor commands finally flow through the ANs that innervate the ipsilateral lateral rectus muscle, when the same innervation is exerted to the contralateral medial rectus muscle by intervention from the abducens internuclear neurons.

A brainstem saccadic circuitry, corroborated by several contributions of local field potentials (LFPs) to the dynamics of neuronal synaptic activity between three neural populations in generating horizontal and vertical saccades in two rhesus monkeys, was introduced (Van Horn et al., 2010). The extracellular recordings, including spike trains and LFPs, were taken from the saccadic burst neurons (SBNs) in the PPRF at the premotor level, the OPNs in the nucleus raphe interpositus, and the motoneurons at the motor level. The stimulus reconstruction technique was used to measure the accuracy of spike trains and LFPs in reconstructing eye velocity from the data—including 40 saccades for each neuron. The corresponding coding fraction metrics revealed that LFPs from each neuron encode the eye velocity in both the ipsilateral and contralateral directions. In addition, LFP response amplitude of SBNs was described as a function of saccade direction (in 400 saccades) by fitting Gaussian curves to data (see Fig. 8B in Van Horn et al., 2010), indicating that the SBN LFPs can be fine-tuned over all the directed saccades. Dynamic analysis of hyperpolarizing LFPs revealed that the encoded velocity signals relate to the inhibitory drives to motoneurons (during contralateral saccades) and to OPNs (during all saccades).

While the midbrain coordination mechanism in generating saccades has been qualitatively studied (Coubard, 2013, Enderle, 1994, Walton et al., 2005), a complete neural circuitry that includes both the premotor and motor neurons in quantifying the final motoneuronal command to eye muscles has not yet been developed. The utility of SNNs to the biophysical modeling of interconnected neurons (Ghosh-Dastidar and Adeli, 2007, 2009, Mohemmed et al., 2012) elucidates broad principles to modeling at higher structural scales, such as the circuit (Enderle and Zhou, 2010, Rossell'o et al., 2009) and the systems levels (Ramanathan et al., 2012). We investigated neural control of three large saccades. A parallel-distributed and hierarchical neural network model of the midbrain was first presented. To develop the quantitative computational models that establish the basis for this functional neural network model, the saccade burst generator dynamics were described. A neural circuit model was later demonstrated and parameterized to match the firing characteristics of eight neuron populations at both the premotor and motor stages of the neural network. Despite the complexity of the saccade generator in a large-scale SNN, our neural modeling approach led us to address the challenges involved in the implementation of the midbrain pathways due to the heavy storage and computational requirements. The ensuing computational cost was reasonable because of the rationalized modular programming of the neural network.

A time-optimal neuronal controller for human saccadic eye movements was first proposed based on experimental data analysis (Clark and Stark, 1975). Exactly how there is a one-to-one relationship between the firing rate in agonist neurons and the saccade magnitude is a matter of controversy in the literature. For reference, firing rate-saccade amplitude dependent controllers were proposed (Gancarz and Grossberg, 1998, Scudder, 1988). These studies lacked the use of a homeomorphic oculomotor plant, and none of the investigated controllers thereof offered the feasibility of a time-optimal control strategy. The 1st-order time-optimal controller, introduced in Section 1.9, is used herein, which includes the activation and deactivation time constants in agonist and antagonist controller models. This controller has been proven to agree with the experimental findings (Clark and Stark, 1975, Enderle and Wolfe, 1988a). The set of agonist-antagonist controllers of the oculomotor plant supports the time-optimal control theory, in that the motoneurons' firing rate does not determine saccade magnitude. It is noteworthy that the duration of agonist burst discharge solely determines the saccade magnitude based on Fig. 2.19. In view of this, the dependence of the three large saccades on the agonist pulse duration has been found to be well presented by the time-optimal controller.

The control simulation results in Fig. 2.20 substantiate the time-optimal controller due to the close congruence between them and the analytical solutions of saccade characteristics. Notably, the saccade duration-saccade magnitude characteristic (see Fig. 1.24) corroborates our simulation results. These observations give rise to the accuracy of the membrane parameters in the modeling of each neuron, listed in Table 2.2. The proposed saccadic circuitry herein is a complete model of saccade generation, since it not only includes the neural circuits at both the premotor and motor stages of the saccade generator, but it also uses a time-optimal controller to stimulate the oculomotor rectus muscles.

2.7   GLISSADES AS ONE OF THE DEFICIENCIES IN THE OCULOMOTOR CONTROL MECHANISM

Glissades are known as one of the post-saccade phenomena that cause short, flimsy transitions toward the end of a saccade. Experimental saccadic data were gathered from three human subjects exhibiting both normal saccades and dynamic overshoots or glissades (Enderle and Wolfe, 1988a). The analytical solutions of position and velocity characteristics using the parameter estimates for all the saccades matched that of the experimental data.

To date, quite a few studies have examined the effects of deficiencies in the oculomotor control mechanism, especially in saccades, vergence, or accommodative eye movements. The visualization of concussive forces within the core brain structures of patients with diffuse traumatic brain injury (DTBI) revealed the presence of focal sheer stresses at control areas of the basal ganglia, corpus callosum, and midbrain (Tyler, 2013). An analysis of abnormalities in the dynamics of vergence eye movements observed from 16 DTBI patients led to the classification of them as weak, slow, and noisy trajectories, contingent on the extent of deviation from the steady-state normal amplitude. Ciuffreda et al. (2007) reported on a number of different oculomotor version dysfunctions, including saccadic and pursuit deficits, saccadic intrusions, and nystagmus, diagnosed in 82 TBI patients. The most frequent type of dysfunction was saccadic deficits in 62 patients (see Table 6 in Ciuffreda et al. (2007)).

The occurrence of glissades in saccades, recorded from a pool of participants reading text images and from a scene perception experiment, has been detected 47.8% and 59.1%, respectively, (Nyström Holmqvist, 2010). A data-driven algorithm using an adaptive velocity threshold is developed for glissade detection under the defined low-velocity and high-velocity constraints. The glissadic duration in this study was found to range from 10 ms–35 ms. This investigation allowed for analyzing the glissades as a distinctive type of eye movements. Larsson et al. (2013) takes a new systematic approach to delineation of physiological and mathematical properties of postsaccadic oscillations (PSO) when dealing with dynamic scenes from an experiential database. An all-pole system function served as the basis, in order to detect the PSO and assess the classification performance thereof. The classification sensitivity criterion of their proposed algorithm outperformed that of Nyström and Holmqvist (2010) in detection of PSO. Three propositions regarding the modeling and event detection of PSO in this study that provide motivation for future research are: 1) the PSO can be present in the saccades, 2) it can be categorized as a unique eye movement, and 3) it can be neglected by replacing it with a simplified 1st-order all-pole model.

Similar to saccades, sequential burst activities in the neural sites of midbrain enact to the generation of glissades (Enderle, 2002, Enderle and Zhou, 2010, Kapoula et al., 1986). The glissades extend the peak velocity vs. magnitude trajectory of the main-sequence diagrams observed from the normal saccades (Enderle and Zhou, 2010, Nyström Holmqvist, 2010). The rest of this chapter describes the glissade dynamics upon which is based the delineation of the neurosensory control of them, in general, and our computational neural modeling thereof, in particular.

2.8   GLISSADE DYNAMICS

Glissades and dynamic overshoot saccades are generated by the antagonist post-inhibitory rebound burst (PIRB) activity, and, in turn, cause a reverse (second) peak velocity. The study of PIRB firing activity is motivated and corroborated by the findings of many investigators. For reference, Jahnsen and Llinas (1984a) explained the rebound burst activities within thalamic neurons subsequent to very pronounced hyperpolarizations. Enderle and Zhou (2010) described the PIRB as a spontaneous, high neural firing at the first 10 ms of EBN activity, with low intensity stimulation. The firing trajectory in the PIRB begins with a rapid rise to a peak firing rate, and gradually descends to a lower steady-state firing rate approximately 10 ms past the onset. The low membrane threshold voltage of the EBN axon hillock is speculated as the biophysical cause in triggering this firing activity. The induced antagonist PIRB activity proved to be the superior hypothesis over the alternative of adjustment of the antagonist duration of burst after coming off the inhibition in generating the dynamic overshoots or glissades in humans. The collected data from three human subjects evidenced that there are more samples of glissades than the normal saccades or with dynamic overshoots. In addition, the data showed that, as saccade magnitude increases, the dynamic overshoot saccades are less frequently observed than those with glissades.

2.8.1   ANALYSIS AND CHARACTERISTICS

The position and velocity trajectory for a 10° glissade are presented in Fig. 2.21. As illustrated, there are two peak velocities characterizing such a saccade. Enderle and Zhou (2010) determined the first peak velocity sensitivity on the agonist muscle parameters, namely, agonist pulse magnitude and duration, and agonist activation time constant. From the established observations, the judicious selection of these parameters led the estimated first peak velocity to match the experimental data. The first peak velocity was found to be mainly dependent upon agonist pulse magnitude, rather than other agonist parameters variations. Glissades and dynamic overshoot saccades—contrary to the normal saccades—show a second peak velocity. Enderle (2002) described that this second peak velocity is attributed to the antagonist PIRB firing, a characteristic present in the post-saccade phenomena. The dependency of the second peak velocity on the antagonist muscle parameters, i.e., antagonist rebound burst magnitude and antagonist activation time constant, was analyzed (Enderle and Zhou, 2010). The adjustment of these parameters for the purpose of providing desired antagonist dynamics showed that the rebound burst magnitude is the key parameter that determines the second peak velocity.

As shown in Fig. 2.21, the glissade starts when the velocity rises from $0° \text{ s}^{-1}$ toward the first peak velocity, and terminates a few milliseconds after the appearance of the second peak velocity. The overshoot during the glissades returns to the steady-state level in a gradual manner. Hence, the second peak velocity is smaller in the glissades than that of the dynamic overshoot saccades. Another disparity between the glissades and dynamic overshoot saccades is that the antagonist rebound burst magnitude is smaller in the glissades than in the dynamic overshoot saccades.

2.8.2   NEURAL CONTROLLER WITH PIRB

As with the saccade burst generator, the neuroanatomical connectivity structure in Fig. 2.1 provides a neural, glissadic pathway at both the premotor and motor levels in eliciting the final motoneuronal command to eye muscles. In view of the neural network dynamics that needs to be treated in the case of spiking neurons, the sequence of neural firing, outlined in Subsection 2.3.3, needs to be executed. The antagonist PIRB is stimulated primarily by the rebound burst activity of the EBN on the contralateral side of the neural network. The stimulation of the contralateral EBN itself originates from the ipsilateral FN.

In order to provoke this rebound burst activity in the sequence of neural firing, one modification should be made in the seventh step: the resumption of tonic firing and PIRB activity in the contralateral AN should not occur until soon after the ipsilateral IBN stops firing. The PIRB lasts as long as the ipsilateral FN continues to burst and spread.

The general applicability of SNNs to the biophysical modeling of interconnected neurons provides efficiency at large scale simulation of spiking neurons. This perspective forms the basis for modeling at different structural scales, such as the circuit and the systems levels, as discussed earlier in this chapter. The glissade-induced spiking activities at the premotor level are modeled with an HH model for the bursting neurons, and with a modified FHN model for the tonic-spiking neurons (Faghih et al., 2012). Consequently, the neural circuitry introduced in Section 2.4 is used to capture the neural modeling at the circuit level. The membrane parameters are as provided in Table 2.2. The circuit model of the neural system for the glissade controller is initially stimulated in the region representing the SC, propagates through the necessary pathways, and ultimately yields an output at the nodes representing the medial and lateral rectus eye muscles. To innervate these muscles by the oculomotor neural activity, rapprochement from a 1st-order time-optimal neural controller is required (see Sections 1.2 and 1.9). Recall that the activation time constant for the brief PIRB is included in the antagonist neural controller described in Eq. (1.30). Shown in Fig. 2.22 are the antagonist neural input and the active-state tension related to a 10° glissade. The PIRB activity starts from T_3 and continues until T_4, with the activation time constant of τ_{tac} in the neural controller during the PIRB interval, and the deactivation time constant τ_{tde} both before, and subsequent to, the PIRB interval.

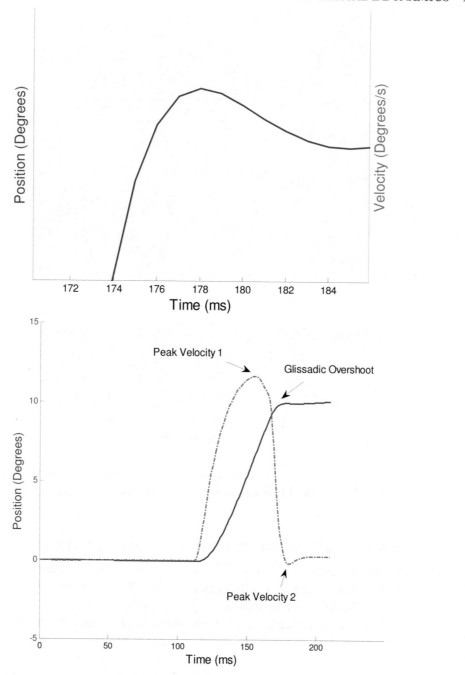

Figure 2.21: Position and velocity trajectories for a typical glissade. The overshoot prior to the settlement at steady-state is magnified. The PIRB activity starts at approximately 160 ms. Evidently, this activity induces a second peak velocity.

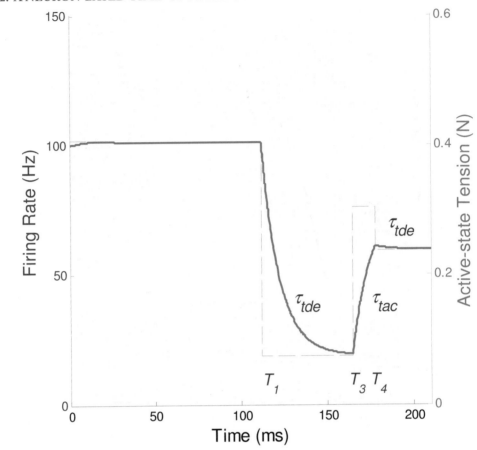

Figure 2.22: Antagonist active-state tension (solid) and neural input (dashed) from the contralateral AN for a sample glissade. The PIRB starts at approximately 160 ms and lasts for 10 ms.

2.8.3   COMPARISON OF GLISSADES TO NORMAL SACCADES

The presentation of control simulation results of three large glissades, namely, 10°, 15°, and 20°, incorporating the neural network controller in a 3rd-order linear homeomorphic muscle model in humans, follows. As elaborated in Section 1.9, the time-optimal neural controller is well suited to the examination of normal saccades or those exhibiting dynamic/glissadic overshoots. As demonstrated by the estimation of neural controller parameters for small and large saccades in Section 1.7, the agonist and antagonist activation time constants are of vital importance in characterizing the dynamics of saccades and glissades. The estimated activation and deactivation time constants illustrated in Fig. 1.28, and the oculomotor plant parameters for humans in Table 1.2, are extracted. As with the normal saccades, the onset delay before glissade, peak firing rate, and burst termination time for the neural populations, are determined according to Table 2.1. Each glissade starts at 120 ms. Table 2.4 indicates the agonist burst duration and the antagonist PIRB duration for all the glissades. The selection of the duration of the agonist burst is congruent with Fig. 1.25B of these parameter estimates, and the duration of the antagonist PIRB conforms with Fig. 1.26B of the duration estimates for large glissades.

Figure 2.23 presents the trajectories of dendritic membrane potential, axonal membrane potential, and synaptic current pulse train for the LLBN, EBN, and AN on both sides of the neural network for a 10° glissade. From the comparison of each trajectory on the ipsilateral side to the contralateral side, it follows that the PIRB activity on the contralateral side is the source of binocular coordination error in the glissades. The brief excitation of the contralateral EBN from the ipsilateral FN leads this neuron to burst at high rates after being released from inhibition by the ipsilateral IBN. The burst trajectory due to the PIRB becomes evident with a rapid rise to a peak firing rate, and gradually descends to a lower steady-state firing rate approximately 10 ms past the onset. When compared to the normal saccades, the induced PIRB firing in the contralateral AN influences the return to a tonic firing rate at steady-state. The timing onset and duration of this PIRB firing in the sequence of neural synaptic connections in Fig. 2.4 necessitated several parameter tunings on the circuitry. It is noteworthy that the delay in synaptic flow for each neuron type agreed with its place within the neural network's hierarchical order at both premotor and motor levels.

The ipsilateral agonist and antagonist burst-tonic firing rates and their corresponding active-state tensions for the glissades are illustrated in Fig. 2.24. Recall that the active-state tensions are low-pass filtered neuronal firing rates based on Eqs. (1.29) and (1.30).

It is noted that the ipsilateral agonist firing rate does not depend on the glissade magnitude, and the duration of the agonist burst solely determines such magnitude based on the physiological constraints. This observation supports the time-optimal control strategy of the oculomotor plant during the large glissades. Moreover, the presented neural stimulation signals are consistent with those provided in Section 1.7 on the basis of the parameter estimates (analytical approach) from the system identification technique. It was found that the agonist activation time constant ascertains the agonist pulse magnitude and, in turn, the first peak velocity. Table 2.4 includes the agonist pulse duration and magnitude from the results in Fig. 2.24. The antagonist controller model parameters contribute significantly to the dynamics of the PIRB. It was our assertion that the PIRB magnitude is the key parameter that ascertains the second peak velocity in glissades. Listed in Table 2.4 are the PIRB duration and magnitude from the demonstrations in Fig. 2.24. These properties are consistent with the parameter estimates of the PIRB magnitude and duration in the antagonist motoneuron for large glissades provided in Fig. 1.26.

The precise dynamics of glissade trajectory is governed by the set of agonist-antagonist controller models, as emphasized previously. The final motoneuronal active-state tensions drive the 3rd-order linear homeomorphic model of the oculomotor plant. Figure 2.25 shows the control simulation results of the movement trajectory for the three glissades and their corresponding

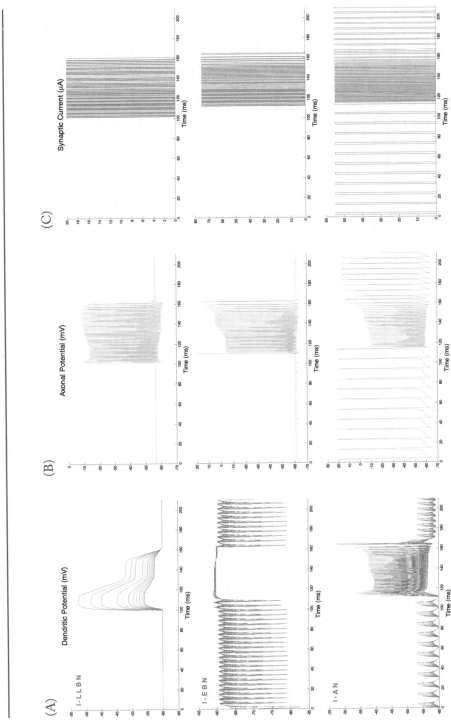

Figure 2.23: The dendritic membrane potential in mV (A), axonal membrane potential in mV (B), and synaptic pulse current train in μA (C) of neurons in a 10° glissade neural controller. Ipsilateral LLBN, EBN, and AN are shown in order. The PIRB activity starts by the stimulation of the contralateral LLBN and EBN for approximately 10 ms (glissade onset: 120 ms).

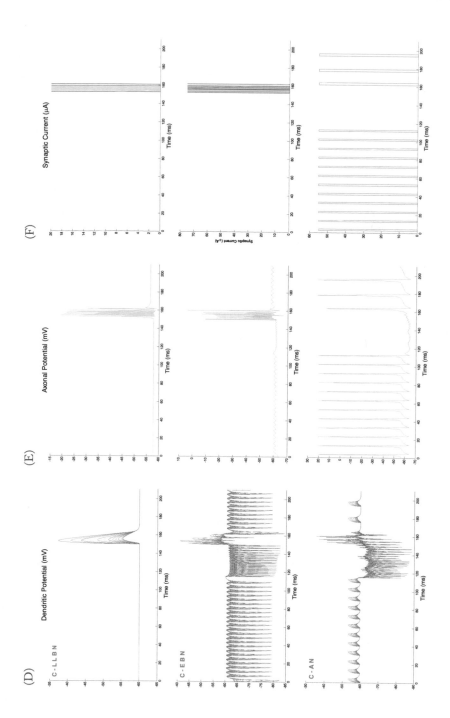

Figure 2.23: *(Continued.)* The dendritic membrane potential in mV (D), axonal membrane potential in mV (E), and synaptic pulse current train in μA (F) of the contralateral LLBN, EBN, and AN in a 10° glissade neural controller. Note the differences between the membrane behavior of each neuron on the ipsilateral side and on the contralateral side.

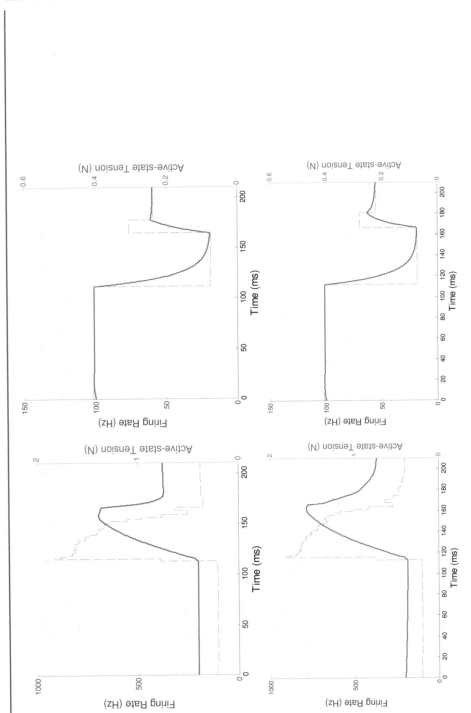

Figure 2.24: The ipsiliteral neural stimulation signals for the agonist and antagonist neural control inputs (dashed), and the corresponding active-state tensions (solid) plotted on the same graph. (Top) 10° glissade and (bottom) 15° glissade. The antagonist controller model accommodates the PIRB activity in the antagonist motoneuron. The agonist and antagonist activation time constants have a key impact on glissade dynamics.

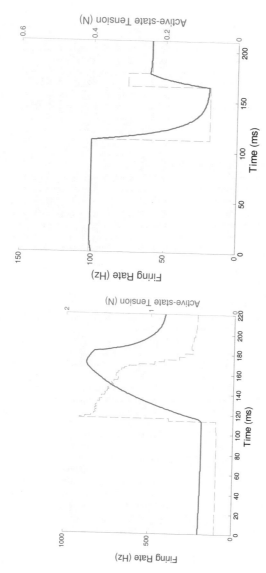

Figure 2.24: *(Continued.)* The ipsiliteral neural stimulation signals for the agonist and antagonist neural control inputs (dashed), and the corresponding active–state tensions (solid) plotted on the same graph in a 20° glissade. The antagonist controller model accommodates the PIRB activity in the antagonist motoneuron. The agonist and antagonist activation time constants have a key impact on glissade dynamics.

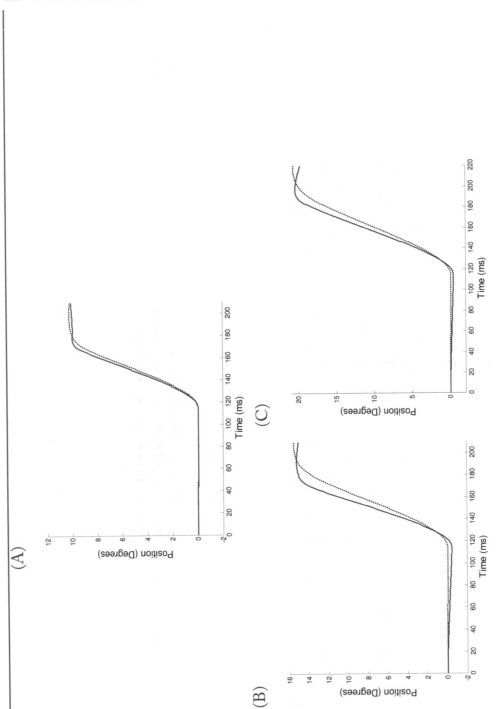

Figure 2.25: The ipsiliteral control simulation results for the eyes undergoing a normal saccade (dotted) and a glissade (solid). (A) 10° displacement, (B) 15° displacement, and (C) 20° displacement. The movement trajectory of glissades deviates notably from that of normal saccades immediately after the PIRB onset in the contralateral AN.

Table 2.4: The agonist and antagonist controller models characterized by the pulse magnitude and duration. The PIRB is exemplified in generating three different glissades

Glissade Magnitude (Degrees)	Agonist Burst Duration (ms)	Agonist Pulse Magnitude (N)	Antagonist PIRB Duration (ms)	Antagonist PIRB Magnitude (N)
10	50	1.39	10	0.24
15	56	1.57	10	0.26
20	65	1.75	10	0.27

normal saccades, depicted in Fig. 2.20. From the comparison of these trajectories, the distinction between the normal saccades and glissades is apparent, due to the PIRB activity in the latter. It is obvious that the discrepancy between the movement trajectories is minimized for the 10° magnitude and increases as the magnitude rises. With the advent of the PIRB activity in the contralateral AN, it is evident that the rectus muscles move the eyeball to steady-state position slower than with the normal saccades. It is interesting, however, that the antagonist neural controller accommodates this variation in the movement trajectory caused by the PIRB. From the velocity profiles of the glissades examined herein, it was gathered that the PIRB elicited a second peak velocity. Comparison of the simulation results herein with the model prediction results in Section 1.7 demonstrated excellent agreement. In either case, the PIRB behavior is well represented by the antagonist neural controller model. In addition, the duration of the agonist burst has the dominant effect on achieving the desired saccade/glissade magnitude consistent with physiological constraints. Analogous to the saccade neural system, the glissade neural system is a complete system, since it not only includes the premotor burst generator, but it also uses a time-optimal controller to yield the glissade magnitude.

Many patients diagnosed with DTBI have revealed eye movement trajectories with losses of binocular coordination, as evidenced by the notable deviations in saccades (Tyler, 2013). It is highlighted that existing clinical brain-imaging protocols do not offer a reliable framework by which to detect the damage to the core brain structures in DTBI. The continuing research effort in the subsumed neurology field has been motivated by the need to discern the physiological limitations of binocular eye movements, and the proper pharmacological and visual therapies for deficits thereof. As an analysis of the presence of a saccade oculomotor abnormality, the dependency of the PIRB magnitude and the duration on the saccade magnitude was determined for a broad variety of the small and large saccades with overshoots (Enderle and Zhou, 2010).

Thiagarajan et al. (2011) reviewed the latest clinical observations and laboratory investigations of a variety of static and dynamic vergence dysfunctions in mild TBI. The statistical tests on the mean values of 14 static parameters indicated that there are considerable differences in five parameters between the mild TBI patients and a group of visually normal individuals. As a neurological implication of this study, the pulse component of a time-optimal controller is the cause of the vergence oculomotor abnormality within the symmetric vergences (convergence and divergence) in mild TBI. Multiple impacts to the helmet shell and facemask of 25 concussed NFL players caused changes in biomechanical parameters, and were explored and analyzed (Viano et al., 2007). The demonstration of principal strain in the midbrain for five concussed players indicated the occurrence of the maximum strains in it, as a response to variations in the head kinematics after collision (see Fig. 2 in Viano et al. (2007)). Clinical examinations of the two group of patients diagnosed with TBI—one from the Palo Alto Polytrauma Rehabilitation Center (PRC) and the other from the Polytrauma Network Site (PNS)—revealed notable binocular vision dysfunctions (Cockerham et al., 2009). Among the visual complaints was the "pursuit/saccade insufficiency" reported to be in 29% of cases in the PRC group and in 23% of cases in the PNS group.

The comparison of glissades with normal saccades herein confirms the fact that glissades are observed because of anomalies in saccade neuronal programming. Our theory was that the deficits are manifested due to an unplanned PIRB firing in the antagonist motoneurons as a source of coordination error in returning to tonic firing rates. The mathematical and computational neural modeling of anomalies in the dynamics of the saccades is insightful because it has the potential of diagnosing neurological disorders such as traumatic brain injury.

Bibliography

Bahill, A. T. (1980) Development, validation and sensitivity analyses of human eye movement models. *CRC Crit. Rev. Bioeng.*, 4 (4): 311–355.

Bahill, A.T., (1981) *Bioengineering: Biomedical, Medical and Clinical Engineering*, Prentice-Hall, Inc., Englewood Cliffs, NJ.

Bahill, A.T. and Hamm, T.M., (1989) Using open-loop experiments to study physiological systems, with examples from the human eye movement systems, *News in Physiol. Sci.*, 4: 104–109.

Bahill, A.T., Latimer, J.R., and Troost, B.T., (1980) Linear homeomorphic model for human movement. *IEEE Trans. Biomed Engr.*, BME-27, 631–639. DOI: 10.1109/TBME.1980.326703. 8, 35, 71, 73, 96

Behan, M. and Kime, N.M., (1996) Intrinsic Circuitry in the Deep Layers of the Cat Superior Colliculus. *Vis. Neurosci.*, 13: 1031–1042. DOI: 10.1017/S0952523800007689.

Bruce, C. J., Goldberg, M. E., Bushnell, M.C., and Stanton, G.B., (1985) Primate frontal eye fields. II. Physiological and anatomical correlates of electrically evoked eye movements. *Journal of Neurophys.*, vol. 54, no. 3: 714–34.

Cannon, S.C. and Robinson, D.A., (1987) Loss of the neural integrator of the oculomotor system from brain stem lesions in monkey. *J. Neurophys.*, 57(5): 1383–1409.

Ciuffreda, K.J., Kapoor, N., Rutner, D., Suchoff, I.B., Han, M.E., Craig, S., (2007) Occurrence of oculomotor dysfunctions in acquired brain injury: a retrospective analysis. *Optometry*, 78 (4): 155–161. DOI: 10.1016/j.optm.2006.11.011. 122

Clark, M.R. and Stark, L., (1975) Time optimal control for human saccadic eye movement. *IEEE Trans. Automat. Contr.*, AC-20: 345–348. DOI: 10.1109/TAC.1975.1100955. 121

Cockerham, G.C., Goodrich, G.L., Weichel, E.D., Orcutt, J.C., Rizzo, J.F., Bower, K.S., Schuchard, R.A., (2009) Eye and visual function in traumatic brain injury. J. Rehabil. Res. Dev., 46 (6): 811–8. 134

Coubard, O. A., (2013) Saccade and vergence eye movements: a review of motor and premotor commands. *European J. of Neurosci.*, 38: 3384–3397. DOI: 10.1111/ejn.12356. 77, 120, 121

Enderle, J.D., Wolfe, J.W., and Yates, J.T., The Linear Homeomorphic Saccadic Eye Movement Model—A Modification. *IEEE Transactions on Biomedical Engineering*, 1984, Vol. BME-31, No. 11: 717–720. 8

Enderle, J.D. and Wolfe, J.W., (1987) Time-Optimal Control of Saccadic Eye Movements. *IEEE Transactions on Biomed. Eng.*, Vol. BME-34, No. 1: 43–55. DOI: 10.1109/TBME.1987.326014. 72, 73

Enderle, J.D. and Wolfe, J.W., (1988a) Frequency Response Analysis of Human Saccadic Eye Movements: Estimation of Stochastic Muscle Forces, *Comp. Bio. Med.*, 18: 195–219. DOI: 10.1016/0010-4825(88)90046-7. 1, 8, 30, 34, 58, 96, 121, 122

Enderle, J.D., (1988b) Observations on Pilot Neurosensory Control Performance During Saccadic Eye Movements, *Aviat., Space, Environ. Med.*, 59: 309–313. 33

Enderle, J.D., (1994) A Physiological Neural Network for Saccadic Eye Movement Control. Air Force Material Command, *Armstrong Laboratory AL/AO-TR-1994–0023*: 48 pages. DOI: 10.1155/2014/406210. 77, 80, 121

Enderle, J.D., and Engelken, E.J. (1995) Simulation of Oculomotor Post-Inhibitory Rebound Burst Firing Using a Hodgkin-Huxley Model of a Neuron. *Biomedical Sciences Instrumentation*, 31: 53–58. 77, 80

Enderle, J.D. and Engelken, E.J., (1996) Effects of Cerebellar Lesions on Saccade Simulations, *Biomed. Sci. Instru.*, 32: 13–22.

Enderle, J.D., (2002) *Neural Control of Saccades*. In J. Hyönä, D. Munoz, W. Heide and R. Radach (Eds.), The Brain's Eyes: Neurobiological and Clinical Aspects to Oculomotor Research, Progress in Brain Research, V. 140, Elsevier, Amsterdam, 21–50. 10, 30, 36, 39, 44, 57, 58, 66, 77, 80, 123

Enderle, J.D., Blanchard, S.M. and Bronzino, J.D., (2005) *Introduction to Biomedical Engineering (Second Edition)*, Elsevier, Amsterdam, 1118 pages. 32

Enderle, J.D., (2006a) *Eye Movements*. In Wiley Encyclopedia of Biomedical Engineering (Metin Akay, ed.) Hoboken: John Wiley & Sons, Inc. DOI: 10.1002/9780471740360.

Enderle, J.D., (2006b) *The Fast Eye Movement Control System*. In: The Biomedical Engineering Handbook Biomedical Engineering Fundamentals, Third Edition, ed. J. Bronzino. CRC Press, Inc., Boca Raton, FL, Chapter 16, pages 16–1 to 16–21.

Enderle, J.D., and Zhou, W., (2010) *Models of Horizontal Eye Movements. Part 2: A 3rd-Order Linear Saccade Model*. Morgan & Claypool Publishers. DOI: 10.2200/S00263ED1V01Y201003BME034. 76, 77, 78, 80, 82, 84, 85, 86, 87, 88, 89, 91, 93, 94, 103, 110, 121, 123, 133

Enderle, J.D., and Bronzino, J.D., (2011) *Introduction to Biomedical Engineering 3rd edn.* Elsevier, Amsterdam. 80, 88

Enderle, J.D., Engelken, E.J., and Stiles, R. N., (1990) Additional Developments in Oculomotor Plant Modeling. *Biomedical Sciences Instrumentation,* 26: 59–66.

Enderle, J.D., Engelken, E.J., and Stiles, R.N. A comparison of static and dynamic characteristics between rectus eye muscle and linear muscle model predictions. (1991) *IEEE Trans. Biomed. Eng.,* 38:1235–1245. DOI: 10.1109/10.137289. 30, 37

Engelken, E.J., Stevens, K.W. and Enderle, J.D., (1991a) Optimization of an Adaptive Non-linear Filter for the Analysis of Nystagmus, *Biomed. Sci. Instru.,* 27: 163- 170.

Engelken, E.J., Stevens, K.W. and Enderle, J.D., (1991b) Relationships between Manual Reaction Time and Saccade Latency in Response to Visual and Auditory Stimuli, *Aviat., Space, Environ. Med.,* 62: 315–318.

Engelken, E.J., Stevens, K.W., McQueen, W.J. and Enderle, J.D., (1996) Application of Robust Data Processing Methods to the Analysis of Eye Movements, *Biomed. Sci. Instru.,* 32: 7–12.

Faghih, R.T., Savla, K., Dahleh, M.A., Brown, E.N., (2012) Broad range of neural dynamics from a time-varying FitzHugh–Nagumo model and its spiking threshold estimation. *IEEE Trans. on Biomed. Eng.,* 59 (3): 816–823. DOI: 10.1109/TBME.2011.2180020. 76, 89, 103, 124

Fuchs, A.F., and Luschei, E.S., (1970) Firing patterns of abducens neurons of alert monkeys in relationship to horizontal eye movement. *J Neurophysiol,* 33 (3), 382–392. 66

Fuchs, A. F. and Luschei, E. S., (1971) Development of isometric tension in simian extraocular muscle. *J. Physiol.* 219, 155–66. 46

Fuchs, A.F., Kaneko, C., Scudder, C. (1985a) Brainstem Control of Saccadic Eye Movements. *Ann Rev. Neurosci,* 8:307–337. DOI: 10.1146/annurev.ne.08.030185.001515. 57, 71

Gancarz, G., and Grossberg, S., (1998) A neural model of the saccade generator in reticular formation. *Neural Networks,* vol. 11: pp. 1159–1174. DOI: 10.1016/S0893-6080(98)00096-3. 82, 83, 121

Gandhi, N.J., and Keller, E.L., (1997) Spatial distribution and discharge characteristics of the superior colliculus neurons antidromically activated from the omnipause region in monkey. *J Neurophysiol,* 76: 2221–5.

Ghosh-Dastidar, S., and Adeli, H., (2007) Improved spiking neural networks for EEG classification and epilepsy and seizure detection. *Integr. Comput.-Aided Eng.,* 14: 187–212. 75, 88, 121

Ghosh-Dastidar, S., and Adeli, H., (2009) Spiking neural networks. *Int. J. of Neural Systems*, 19 (4), 295–308. DOI: 10.1142/S0129065709002002. 75, 88, 121

Girard, B., and Berthoz, A., (2005) From brainstem to cortex: Computational models of saccade generation circuitry. *Progress in Neurobiology*, 77: 215–251. DOI: 10.1016/j.pneurobio.2005.11.001.

Goldstein, H. (1983) The neural encoding of saccades in the rhesus monkey (Ph.D. dissertation). Baltimore, MD: The Johns Hopkins University. 1, 2, 8

Goldstein, H.P., and Robinson, D.A., (1986) Hysteresis and slow drift in abducens unit activity. *Journal of Neurophysiology*, vol. 55, pp. 1044–1056.

Harris, C.M., and Wolpert, D.M. (2006) The main sequence of saccades optimizes speed-accuracy trade-off. *Biol Cybern*, 95 (1), 21–29. DOI: 10.1007/s00422-006-0064-x. 73

Hikosaka, O. and Wurtz, R.H., (1983a) Visual and oculomotor functions of monkey substantia nigra pars reticulata. I. Relation of visual and auditory responses to saccades, *J. Neurophys.*, 49 (5): 1230–53.

Hikosaka, O. and Wurtz, R.H., (1983b) Visual and oculomotor functions of monkey substantia nigra pars reticulata. II. Visual responses related to fixation of gaze, *J. Neurophys.*, May 49(5): 1254–67.

Hodgkin, A. L. and Huxley, A. F., (1952) A quantitative description of membrane current and its application to conduction and excitation in nerve. *J. Physiol.* 117:500. DOI: 10.1007/BF02459568. 76

Hodgkin, A.L., Huxley, A.F., and Katz, B., (1952) Measurement of Current-Voltage Relations in the Membrane of the Giant Axon of *Loligo*. *Journal of Physiology*, 116: 424–448.

Hu, X., Jiang, H., Gu, C., Li, C., Sparks, D. (2007) Reliability of Oculomotor Command Signals Carried by Individual Neurons. *PNAS*, 8137–8142. DOI: 10.1073/pnas.0702799104. 57, 58, 73

Izhikevich, E. M. (2003) Simple model of spiking neurons. *IEEE Trans. Neural Netw.*, 14: 1569–1572. DOI: 10.1109/TNN.2003.820440. 76

Jahnsen, H. and Llinas, R., (1984a) Electrophysiological properties of guinea pig thalamic neurones: an in vitro study. *J. Physiol.*, London, 349: 205–226. 123

Jahnsen, H. and Llinas, R., (1984b) Ionic basis for the electroresponsiveness and oscillatory properties of guinea pig thalamic neurons in vitro. *J. Physiol.*, London, 349: 227–247.

Kandel, E. R., Schwartz, J. H., and Jessell, T. M., (2000) *Principles of Neural Science: Fourth Edition*. McGraw-Hill Co., New York.

Kapoula, Z., Robinson, D.A. and Hain, T.C., (1986) Motion of the eye immediately after a saccade. *Exp. Brain Res.*, 61: 386–394. DOI: 10.1007/BF00239527. 123

Keller, E.L., McPeek, R.M. and Salz, T., (2000) Evidence against direct connections to PPRF EBNs from SC in the monkey. *J. Neurophys.*, 84 (3): 1303–13.

Krauzlis, R.J., (2005) The control of voluntary eye movements: new perspectives. *The Neuroscientist*, 11 (2):124–137. DOI: 10.1177/1073858404271196.

Larsson, L., Nyström, M., and Stridh, M., (2013). Detection of Saccades and Postsaccadic Oscillations in the Presence of Smooth Pursuit. *IEEE Trans. on Biomed. Eng.*, 60 (9): 2484–2493. DOI: 10.1109/TBME.2013.2258918. 122

Leigh, R. J., and Zee, D. S., (1999) The Neurology of Eye Movements. *Oxford University Press Inc.*, New York, New York. DOI: 10.1097/WNO.0b013e3180334d7e.

Ling, L., Fuchs, A., Siebold, C., and Dean, P., (2007) Effects of initial eye position on saccade-related behavior of abducens nucleus neurons in the primate. *J Neurophysiol*, 98 (6), 3581–3599. DOI: 10.1152/jn.00992.2007. 66

Miura, K., Optican, L., (2006) Membrane Chanel Properties of Premotor Excitatory Burst Neurons May Underlie Saccade Slowing After lesions of Ominpause Neurons. *J. Comput Neurosci.*, 20: 25–41. DOI: 10.1007/s10827-006-4258-y.

Mohemmed, A., Schliebs, S., Matsuda, S., and Kasabov N., (2012) SPAN: Spike Pattern Association Neuron for learning spatio-temporal spike patterns. *Int. J. of Neural Systems*, 22 (4): 1–16. DOI: 10.1142/S0129065712500128. 75, 88, 121

Munoz D. P. and Wurtz, R. H., (1995) Saccade-related activity in monkey superior colliculus. I. Characteristics of burst and buildup cells, *J. Neurophysiol.*, 73: 2313–2333.

Munoz, D.P. and Istvan, P.J., (1998) Lateral inhibitory interactions in the intermediate layers of the monkey superior colliculus, *J. Neurophys.*, Mar 79(3): 1193–209.

Nakahara, H., Morita, K., Wurtz, R.H., and Optican, L.M. (2006) Saccade-Related Spread of Activity Across Superior Colliculus May Arise From Asymmetry of Internal Connections. *J Neurophysiol.*, 96: 765–774. DOI: 10.1152/jn.01372.2005.

Nyström, M. and Holmqvist , K., (2010). An adaptive algorithm for fixation, saccade, and glissade detection in eyetracking data. *Behav. Res. Methods*, 42 (1): 188–204. DOI: 10.3758/BRM.42.1.188. 122, 123

Optican, L.M., and Miles, F.A. (1985) Visually induced adaptive changes in primate saccadic oculomotor control signals. *J Neurophysiol*, 54 (4), 940–958. 2

Quaia, C. and Optican, L.M. (2003) Dynamic eye plant models and the control of eye movements. *Strabismus, 11* (1), 17–31. DOI: 10.1076/stra.11.1.17.14088.

Ramanathan, K., Ning, N., Dhanasekar, D., Guoqi, L., Luping, S., and Vadakkepat, P. (2012) Presynaptic learning and memory with a persistent firing neuron and a habituating synapse: A model of short term persistent habituation. *Int. J. of Neural Systems*, 22(4): 1250015. DOI: 10.1142/S0129065712500153. 76, 88, 121

Ramat, S., Leigh, R.J., Zee, D., Optican, L. (2005) Ocular Oscillations generated by Coupling of Brainstem Excitatory and Inhibitory Saccadic Burst Neurons. *Exp. Brain Res*, 160: 89–106. DOI: 10.1007/s00221-004-1989-8.

Robinson, D.A., (1973) Models of the saccadic eye movement control system. *Kybernetik*, 14: 71–83. DOI: 10.1007/BF00288906.

Robinson, D.A. (1981) Models of mechanics of eye movements. In: B.L. Zuber (Ed.) *Models of Oculomotor Behavior and Control* (pp. 21–41). Boca Raton, FL: CRC Press. 8, 30, 43, 66, 72, 82, 83

Rosselló, J.L., Canals, V., Morro, A., and Verd, J., (2009) Chaos-based mixed signal implementation of spiking neurons. *Int. J. of Neural Systems*, 19 (6): 465–471. DOI: 10.1142/S0129065709002166. 75, 88, 121

Sato, H. and Noda, H., (1992) Saccadic dysmetria induced by transient functional decoration of the cerebellar vermis, *Brain Res. Rev.*, 8 (2): 455–458. DOI: 10.1007/BF02259122.

Scudder, C.A., (1988) A new local feedback model of the saccadic burst generator. *J. Neurophysiol.*, 59(5): 1455–1475. 121

Scudder, C.A., Kaneko, C., Fuchs, A., (2002) The brainstem burst generator for saccadic eye movements: a modern synthesis. *Exp. Brain Res.*, 142: 439–462. DOI: 10.1007/s00221-001-0912-9. 120

Short, S.J. and Enderle, J.D., (2001) A Model of the Internal Control System Within the Superior Colliculus, *Biomed. Sci. Instru.*, 37: 349–354.

Sparks, D. L., Nelson, J. S., (1987) Sensory and motor maps in mammalian superior colliculus. *TINS*, vol. 10, no. 8:312–317. DOI: 10.1016/0166-2236(87)90085-3.

Sparks, D.L. and Hartwich-Young, R., (1989) The Deep Layers of the Superior Colliculus. *Rev. Oculomot. Res.*, 3: 213–55.

Sparks, D.L., (2002) The Brainstem Control of Saccadice Eye Movements. *Neuroscience*, 3: 952–964. DOI: 10.1038/nrn986.

Sparks, D.L., Holland, R. and Guthrie, B.L., (1976) Size and Distribution of Movement Fields in the Monkey Superior Colliculus. *Brain Res.*, 113: 21–34. DOI: 10.1016/0006-8993(76)90003-2. 9, 50, 115

Sylvestre, P.A. and Cullen, K.E., (2006) Premotor Correlates of Integrated Feedback Control for Eye–Head Gaze Shifts. *J Neurophysiol,* 26 (18): 4922– 4929. DOI: 10.1523/JNEUROSCI.4099-05.2006. 66

Thiagarajan, P., Ciuffreda, K.J., and Ludlam, D.P., (2011) Vergence dysfunction in mild traumatic brain injury (mTBI): a review. *Ophthalmic Physiol Opt.* 31 (5): 456–68. DOI: 10.1111/j.1475-1313.2011.00831.x. 133

Tyler, C.W., (2013) Binocular eye movements in health and disease. *Proc. of the SPIE,* 8651, id. 86510Y, 15 pp. DOI: 10.1117/12.2012253. 122, 133

Van Gisbergen, J.A., Robinson, D.A., and Gielen, S., (1981) A quantitative analysis of generation of saccadic eye movements by burst neurons. *J Neurophysiol, 45* (3), 417–442. 8, 66, 72

Van Horn, M.R., Mitchell, D.E., Massot, C., and Cullen, K.E., (2010) Local neural processing and the generation of dynamic motor commands within the saccadic premotor network. *J Neurosci.,* 30 (32): 10905–10917. DOI: 10.1523/JNEUROSCI.0393-10.2010. 120, 121

Viano, D.C., Casson, I.R., Pellman, E.J., (2207) Concussion in professional football: biomechanics of the struck player–part 14. Neurosurgery, 61 (2): 313–27. DOI: 10.1227/01.NEU.0000279969.02685.D0. 134

Walton, M.M.G., Sparks, D.L., and Gandhi N.J., (2005) Simulations of saccade curvature by models that place superior colliculus upstream from the local feedback loop. *J. of Neurophysiol.,* 93: 2354–2358. DOI: 10.1152/jn.01199.2004. 121

Westheimer, G., (1954) Mechanism of saccadic eye movements. *AMA Archives of Ophthalmology,* 52: 710–724. DOI: 10.1001/archopht.1954.00920050716006.

Widrick, J.J., Romatowski, J.G., Karhaneck, M., and Fitts, R.H., (1997) Contractile properties of rat, rhesus monkey, and human type I muscle fibers. *Am. J. Physiol.,* 272, R34–42. 48

Wong, W.K., Wang, Z., and Zhen, B., (2012) Relationship between applicability of current-based synapses and uniformity of firing patterns. *Int. J. of Neural Systems,* 22 (4): 1250017. DOI: 10.1142/S0129065712500177. 94, 103

Zee, D.S., Fitzgibbon, E.J. and Optican, L.M. (1992) Saccade-vergence interactions in humans. *J. Neurophysiol.,* 68: 1624–1641. 120

Zhou, W., Chen, X., and Enderle, J. (2009) An Updated Time-Optimal 3rd-Order Linear Saccadic Eye Plant Model. *International Journal of Neural Systems,* Vol. 19, No. 5, 309–330, 2009. DOI: 10.1142/S0129065709002051. 1, 10, 30, 77, 80, 82, 96, 114

Authors' Biographies

ALIREZA GHAHARI

Alireza Ghahari received his B.Sc. degree in electrical engineering from the Sharif University of Technology, Iran, in August 2007. Thereafter, he completed his M.Sc. in electrical and computer engineering at the University of Tehran, Iran, in March 2010. During those years of study, he gained valuable insights into systems engineering by taking courses in a variety of contexts, such as statistical signal processing, information theory and coding, computer vision, and pattern recognition. He started his Ph.D. in the ECE department at University of Connecticut in Fall 2010. His dissertation major advisor was Prof. John Enderle. He has come to realize the profound contributions of Prof. Enderle in the field of theoretical and computational neuroscience, and truly considers John to be a major influence, both academically and personally. His research areas of interest include spiking neural networks analysis and implementation, brain-computer interface, and development of computational techniques.

JOHN D. ENDERLE

John D. Enderle is a Professor of Biomedical Engineering and Electrical & Computer Engineering at the University of Connecticut, where he was Biomedical Engineering Program Director from 1997–2010. He received his B.S., M.E., and Ph.D. degrees in biomedical engineering, and M.E. degree in electrical engineering from Rensselaer Polytechnic Institute, Troy, New York, in 1975, 1977, 1980, and 1978, respectively.

Dr. Enderle is a Fellow of the IEEE, the past Editor-in-Chief of the *EMB Magazine* (2002–2008), the 2004 EMBS Service Award Recipient, Past-President of the IEEE-EMBS, and was EMBS Conference Chair for the 22nd Annual International Conference of the IEEE EMBS and World Congress on Medical Physics and Biomedical Engineering in 2000. He is also a Fellow of the American Institute for Medical and Biological Engineering (AIMBE), Fellow of the American Society for Engineering Education, Fellow of the Biomedical Engineering Society, and a Rensselaer Alumni Association Fellow. Enderle is a former member of the ABET Engineering Accreditation Commission (2004–2009). In 2007, Enderle received the ASEE National Fred Merryfield Design Award. He is also a Teaching Fellow at the University of Connecticut since 1998. Enderle is the Biomedical Engineering Book Series Editor for Morgan & Claypool Publishers.

Enderle is also involved with research to aid persons with disabilities. He is the Editor of the NSF book series on NSF Engineering Senior Design Projects to Aid Persons with Disabilities, published annually since 1989. Enderle is also an author of the book *Introduction to Biomedical Engineering*, published by Elsevier in 2000 (first edition), 2005 (second edition), and 2011 (third edition). Over his career, Enderle has been an author of over 200 publications and 49 books or book chapters. Enderle's current research interest involves characterizing the neurosensory control of the human visual and auditory system.